Happy Father's D[illegible]

Pete

June 1992.

Love

Hilary &

Nancy.

GOATS and GOATKEEPING

Katie Thear

MEREHURST PRESS
LONDON

Published 1988 by Merehurst Press
5 Great James Street
London WC1N 3DA

Co-published in Australia and New Zealand by
Child & Associates, Unit C, 5 Skyline Place,
Frenchs Forest, NSW 2086, Australia

ISBN 0 948075 91 0

Editor: Edward Bunting
Designer: Carole Perks
Typeset by Filmset
Colour separation by Fotographics Ltd, London-Hong Kong
Printed in Portugal by Printer Portuguesa

Cover photo: Tony Stone Worldwide (Graeme Harris).
Pencil drawings by Graham Turner.
Photographs inside book: AEG Fotostock/18; Heather Angel/47 bottom; Ardea London Ltd (J.B. & S. Bottomley)/39; Stephen Austin Newspapers Ltd/138; E. Betts/14; Biofotos (Murray Watson)/35 top; BMV Picturebank/19,27; Stephanie Colasanti/22 bottom, 26 top, 31 top; Bruce Coleman Ltd (Mark Boulton)/42 top, (Jane Burton)/23, 30, 34, 38 top, 131 top, (J. Canalosi)/47 top, (Hans Reinhard)/26 bottom, (Rod Williams)/43; Dennis Davis/6; Dorset Echo/88; Farmer's Weekly (Peter Adams)/126, (Picture Library)/101 bottom, 113 top, 125, (Charles Topham)/116, 135; Fullwood & Bland Ltd/96 bottom; Claire Housden/38 bottom; Keystone Press Agency/134; Midland Safari Park (Mr & Mrs R.R. Lawrence)/141 top and bottom; Pauline Moss/54; P. Norwood/35 bottom; Anna Oakford/20, 22 top, 36, 46, 50 bottom, 71, 74 bottom, 89, 91, 93, 94, 98, 99 (both), 140, 144, 145 (both), 146, 149, 152, 155, 156, 157; Russell Fine Art/49, 50 top, 136; Chris Some Photography/66; Katie Thear/31 bottom, 42 bottom; Tropix Photo Library (E. Williams)/ 60, 78, 104; Tweed Cashmere Co. (Jasmin Moore)/24, 122; Wheeler Cheese Presses (John Sculpher)/113 bottom; Derek White/52, 139.

CONTENTS

A feral Welsh male goat in the mountains of Snowdonia in North Wales

PREFACE

I have visited goatkeepers and goat enterprises in many parts of the world, including Britain, the USA, Australia, France, Switzerland, Spain and the Netherlands. One aspect which has struck me more than anything else is that the goat world is more alike than it is different.

Goats have a common ancestry, and many of the well-known breeds are to be found all over the world. Systems of management are similar, with obvious climatic variations.

Yet there is a curious tendency to be parochial and to believe that what is happening in one's own country is in some way exclusive. At a time when importations and exportations of goat breeds are increasing, and when international travel generally is becoming more common, it is appropriate to take a more global view.

Within the world of goats there are many scales of enterprise, from the family with a couple of pet goats to the full-time commercial goat farm. Often there is a considerable overlapping of activities: a commercial enterprise may be involved in showing, or owners of pet goats might sell dairy produce. It is a mistake to try and make rigid distinctions in a field where they do not exist.

Both of these considerations have been taken into account in writing this book. If it is to cover the whole topic of 'goats and goatkeeping', as the title claims, it must cater for all those who have an interest in goats, in whatever part of the world they live and on whatever scale they are operating.

I would like to thank all those goatkeepers who have extended their hospitality to me in different parts of the world. I hope that they will understand my decision to dedicate this book, not to them, but to my favourite goat, Florence, a British Saanen who taught me more about goats than anyone else.

KATIE THEAR
WIDDINGTON, 1987

1. INTRODUCTION

Goats have a long and chequered history. They have been closely associated with man since Paleolithic times. The caves at Charente in France bear testimony to this relationship, with bas-relief carvings dating back 16,000 years. There is no evidence that they were domesticated until Neolithic times, over 10,000 years later, when archaeological evidence indicates that they shared man's settlements. The ancient Egyptians kept them, they were common in Biblical times and there are few societies which have not been involved with goats to some degree. This does not mean that they have always been popular. As a species, they have been maligned and, on occasion, persecuted.

Goats in History

The enclosure movement in seventeenth- to nineteenth-century England and Wales deprived large numbers of cottagers of their grazing rights on common land. Although the long-term effect was to increase overall agricultural production, the landless peasantry lost their main source of milk and dairy produce as goats and other livestock were turned off the commons.

In Scotland the Highland clearances had a similar effect; where the crofters were usurped, their goats were also displaced. Many of these animals reverted to the wild, either forming new herds of wandering, feral goats or adding to those already in existence.

Sheep farmers and landowners were soon to wage a war of attrition against feral goats. The sheep farmers suspected goats of interbreeding with their flocks, while landowners looked on the goats' ability to encroach upon enclosed land as tantamount to a revolt of the lower orders.

Britain still has a legacy of this attitude today. Until recently, the farming community has given little credence to the goat as a serious farm animal, while the attitude that the goat is 'the tinker's cow' is still prevalent.

An important stage in the history of goats in Britain was the formation of The British Goat Society (BGS) in 1879. Its aims were to counteract the general prejudice against goats and to spread more accurate information about them. It encouraged cottagers to keep them again and to set about improving and developing the dairying qualities of goats.

Since then, it has organized many schemes which have helped to transform goatkeeping in Britain. The BGS registers and records recognized breeds of goats on behalf of its members and those of affiliated societies. It records transfers of ownership, licences judges for recognized shows and operates an earmarking scheme for stock identification.

A valuable service is the milk sampling and butterfat count scheme which the society instigated with the help of the Milk Marketing Board. The most recent innovation has been the CAE herd monitoring scheme which tests goats to ensure their freedom from the virus infection Caprine Arthritis Encephalitis.

Other countries have similar organizations, such as the American Dairy Goat Association, the Goat Breeders' Society of Australia and the New Zealand Goat Breeders' Society. Most local goat clubs are affiliated to the national society of their respective countries.

The goat's changing status

In recent years, new factors have emerged which are helping to improve the status and popularity of goats. An increasing number of people are moving out of towns to buy small acreages in the countryside. These newcomers tend to work their land as a part-time activity while they derive their main income from another source. The goat fits in well with such an enterprise, providing milk and dairy produce for the family, and possibly a surplus for sale.

Another factor is the increasing public awareness of health in general, and diet in particular. Goats' milk is becoming popular because of its relatively low fat content. It is more digestible because of its smaller fat particles. It also appears to cause fewer reactions in people who tend to get asthma, eczema or other allergic conditions from cow's milk.

It is a paradox that while part-time farming is booming, and resettlement of the countryside is providing the first reversal of rural depopulation since the industrial revolution, many larger commercial farms are declining.

This phenomenon is occurring all over the western world. Gross over-production, leading to cereal and butter 'mountains' or milk and wine 'lakes' is increasingly seen as economically, environmentally and morally unacceptable.

Moves are being made to persuade farmers to diversify their activities into new areas, either by the introduction of quotas to restrict their production, or by incentives such as grants to start new enterprises.

One such possibility is goat farming, particularly in the production of quality products such as milk, cheese, yoghurt and the luxury fibres mohair, cashmere and cashgora. As these aspects have developed, new organizations with a greater emphasis on marketing have appeared. The Goat Producers' Association and the British Angora Society have their equivalents in most countries of the developed world.

Meat and fibre

Most people in our society think of goats as dairy animals; but on a world scale they are primarily meat and hide animals, with milk as an added bonus. They are important in the economy of third world countries because they can thrive in relatively barren terrain, and they have long been a source of sustenance for the poor. There is currently a market for goat meat in developed nations such as New Zealand and Australia, though it is still relatively small.

A potentially lucrative area of goat farming is in the production and marketing of goat fibres such as mohair, cashmere and cashgora (the latter is a cross between the first two). Turkey, South Africa, Lesotho, the USA, Argentina, Australia and New Zealand all have developing mohair industries, and of these South Africa, Texas and Turkey are the major producers.

Most cashmere currently comes from China, Mongolia, Iran, Afghanistan and Pakistan, with China as the main producer.

Britain, Spain and other European countries are beginning to develop their own fibre enterprises for the production of all three fibres.

Goatkeepers large and small

So, there is a great deal happening in the developing world of goats, but it is important not to lose sight of the fact that goat enterprises operate at every scale, from the small domestic goatkeeper to the large-scale commercial operation. Whatever the scale, the basic standards of good husbandry are the same. Goats are intelligent and sensitive creatures, quick to respond to individual attention and affection. They do best in relatively small herds and are ideal for the small family farm.

In my view, they do not respond well, or perform well, in an intensive situation. If they are to be in a large farm situation, the best environment is a hill farm or other marginal land enterprise where reasonably unrestricted browsing is available.

Unlike sheep, they browse on weeds as well as grass; they do well on poor, marginal land; and they improve the grazing by keeping down weeds and shrubby growth. In fact, recent research by the Hill Farming Research Organization in Scotland has established that goats have considerable potential for exploiting marginal land. They are able to improve it and make it suitable for sheep, or use it in their own right as milk, meat or fibre producers.

Goats as family pets

Many people who keep goats do so, not for any commercial reason, but because they like them. The goats may be providers of the family milk supply, or a source of fibres for home spinning, but they are often also family pets. This is an important aspect of goatkeeping, and has been borne in mind throughout the writing of this book.

Goats are some of my favourite animals. They are small enough for children to be able to milk, which explains why they are so popular in farm clubs and in school rural study departments. They are popular inhabitants of city farms and children's zoos. They are amenable and versatile. They are gregarious and friendly. They are often contrary, but life is never dull when there are goats around.

2. ABOUT THE GOAT

If we 'place' the goat in its position relative to the rest of the animal kingdom, it is in the **Artiodactyla** or 'even-toed' order of mammals.

This is a group which includes all the ruminants, plus the pigs and hippopotamuses. (Ruminants are animals which crop herbage quickly, partly chew and swallow it, then store it in a special area of the stomach called a rumen. The food is then regurgitated and chewed at leisure, a process called 'chewing the cud': see page 64.)

Within this order there are six families. Goats are in the **Bovidae**, or 'hollow-horned' family. Other members of this family are cattle, sheep, oxen, buffalo, bison, antelope, yaks and gnus. They are all creatures which have permanent horns, rather than those which are shed, such as those of the deer. (Some goats are born naturally polled, or hornless, but this is a genetic mutation from the norm.)

The subdivision or genus of the Bovidae family to which goats belong is *Capra*. The main species of goats within the group are *Capra hircus*, *C. pyrenaica*, *C. caucasia* and *C. falconeri*.

C. hircus is the direct ancestor of the domestic goat. Known also as the bezoar or pasan, it is found in the wild from Greece to Pakistan. There are two subspecies, *C. hircus aegagrus*, the wild Persian goat, and *C. hircus blythi*, known also as the wild Sind goat of Afghanistan. The Persian goat, *C.h. aegagrus*, is regarded as being the first breed to have been domesticated.

Caprine qualities

Whatever their origins, goats share the same basic characteristics. A knowledge of these allows for a better standard of management. It makes sense to try and provide conditions to suit the goats, as far as this is possible, rather than expect the goats to adapt to an unsuitable environment.

They are essentially creatures of the heights, adapted to leaping and scrambling from one precipitous ledge to another. Their feet are beautifully adapted for this, with small hooves able to draw close together and balance on tiny areas.

Years ago, as a child in Wales, I used to wonder at the stories which told of the wild mountain goats being able to stand on a space no bigger than an old penny. In particularly hard winters, they would descend and raid people's gardens before reverting to the mountainous heights around the Blaenau Ffestiniog area of Gwynedd.

I am not suggesting that it is necessary to provide your goats with mountains, but where they are kept in restricted areas, such as yards, it is a good idea to provide them with a scrambling area. This need only be a group of logs placed at different heights, or a few paving stones. Kids and goatlings love them, and even older, more dignified matrons have been known to cast aside their dignity in favour of more capricious behaviour.

Goats are capricious! Even their most ardent admirers have to admit it. They are intelligent and playful, but they are also devious and contrary. It has been said that goats have slit eyes to enable them to see round corners where the grass is greener.

They are experts at opening gates and finding small gaps through which to squeeze. Washing on the line is an irresistible temptation, while fruit trees are there to be debarked and decimated.

The ability of goats to stand on their hind legs and perform intricate ballet steps in order to strip the top branches of their greenery is legendary. Effective confinement is essential if the relationship is to be a happy one, but the confined area needs to include plenty of interest for the goats.

I have already mentioned a scrambling area. The other important necessity is roughage, in the form of cut branches, hedge prunings and general twiggy growth, placed in hayracks. The cuttings must not be from poisonous plants such as yew or rhubarb leaves, nor should they be from areas that have been sprayed.

The four-part stomach

The stomach of a goat is divided into four distinct areas (page 64) enabling it to consume large amounts of vegetation and digest it over a period of time. This system must have evolved to help ensure survival of the wild goat: like other ruminants, it could snatch herbage where available, then escape the possible attention of predatory carnivores to chew its food at leisure and in safety.

The ruminant goat requires a large proportion of roughage in its diet. How much is a matter of controversy, a legacy perhaps of the lack of research into goats up until recent years.

Browsing behaviour

The natural behaviour of goats in the wild is to eat scrubby material at a certain height. This is thought to give them some protection from being infested with internal parasites, which are usually found in greater concentrations at ground level. You can observe part of this behavioural pattern when you feed hay to domestic goats; if any should fall from the hayrack on to the floor, they will not eat it.

For most goats, the main source of roughage is hay, and this should be available every day. Commercial herds are unlikely to have access to branches and hedge prunings, simply because the management system will not be geared to providing them. Smaller enterprises will rely on this much more, not only because a greater degree of individual attention is possible, but because it keeps down feeding costs. However, even the smallest goat enterprise should not stint on good quality hay.

Whatever management system is followed, it is important to remember that if goats are forced to graze on low-growing plants such as grass because extensive browse material is not available, they will be much more at risk from parasitic infection. In fact, the stall-fed goat which is able to exercise only in a concrete yard is usually much freer of worm burden than those which have access to pasture.

Need for companionship

The goat is a herd animal, gregarious in its outlook. It is not a good idea to keep a solitary goat, although there are cases where this has happened. A single animal will transfer its attentions to its owner to the extent of wanting to be with him or her the whole time.

Perhaps there are people who are prepared to devote themselves to being a full-time goat companion, but most of us would have to acknowledge that there is more to life than goats.

There are instances of a single goat living quite happily with a donkey, and in one case, of an inseparable relationship between a goat and a cat. As a general rule, however, it is better to avoid keeping a single animal.

Dislike of rain

Goats are generally hardy creatures yet they have a common dislike of rain, rapidly seeking shelter when the first drops arrive. This may have originated as a safety device during the evolution of goats, preventing them from slipping on the wet rocks of the craggy heights. It is an important factor to bear in mind, remembering to make shelter available for any animals which may be out grazing and at the mercy of sudden showers.

Shelters should also be available when the sun is particularly strong. White goats have been known to suffer from sunburn, and in tropical areas they are more likely to develop skin cancers than coloured goats.

Need for foot care

Goats' feet have nails which grow continuously. In the wild, the effect of scrambling on rocks has the effect of keeping them worn down and short. In domesticated situations, this is not so, and the feet must be trimmed on a regular basis. The provision of a scrambling area, particularly one which incorporates rocks, will help to keep foot trimming to a minimum, but does not usually do away with it entirely. Further details of this are given on page 144.

These, then, are the main characteristics of goats. A little forethought and effort made to provide them with conditions to their liking will help to produce contented, productive animals with a minimum of problems.

3. BREEDS

As far as we know, the domesticated goat is descended mainly from the wild Persian goat, *Capra hircus aegagrus*; and it was first domesticated around 7000–6000 BC on the slopes of the Zagros mountains in what is now western Iran. Since then, much interbreeding has taken place and localized strains have emerged. Most of the selective breeding, however, has been comparatively recent, taking place during the last 150 years.

The following are the main breeds and types kept domestically or commercially in the developed world. Although I am a British writer, I have tried to avoid making this book too parochial by mentioning only those breeds which are commonly recognized in these islands.

Goats are international, not only in their shared ancestry and development, but in the importing and exporting of stock which is taking place, and which is likely to increase dramatically in future years.

Saanen

The all-white **Saanen** originated in the Saane valley in Switzerland. It has been exported to so many countries that it is now the most widely available of the developed breeds. In Switzerland, it is the goat with the best level of milk production. In Britain, where it was introduced in 1903, 1922 and 1965, it is smaller, with a considerably smaller milk yield, the result mainly of insufficient pure lines to bring about productive selective breeding. Registered Saa-

Saanen male and female champions at the 1987 Saanen Breed Show, Stoneleigh, England

nens in Britain are maintained as a closed breed; and the same situation prevails in the USA.

The **British Saanen** is bigger and has a larger milk yield, making it the most widely kept breed in the UK. The improvement was achieved by crossing Swiss and Dutch Saanens with indigenous British goats, although to be eligible for inclusion in the Herd Book the females must be seven-eighths pure-bred. A similar ruling is enforced in the USA where the **American Saanen** has been developed by upgrading with existing American goats.

In Germany, local goats were upgraded with Saanens imported in 1892, to produce the **German Improved White** (Weisse Deutsche Edelziege). In France, the **French Saanen** has been developed using local bloodlines as well as those imported from Switzerland, Britain, Holland and Germany.

Israel has its **Israeli Saanen** produced by crossing the Mamber with Saanens from Switzerland and the Netherlands. The latter is the home of the **Dutch White** goat, while Belgium has the **Belgian White** or Campine. Czechoslovakia has its own version of the Saanen in the **White Hornless Short-eared** goat.

Romania has the **Banat White**, a blend of local goats, Saanens and the German Improved White. Poland's **Polish Improved White** is also derived from the German version. The USSR has the **Improved North Russian**, developed from its existing North Russian goat upgraded with Saanen imports at the turn of the century. Bulgaria has its own version in the **Bulgarian White Milk** goat.

In Australia, Saanens were also imported at the beginning of the century, after the Second World War, and up until 1959 when a ban on the importation of ruminants was introduced. Local goats were upgraded with the importations to produce some of the finest milking goats in the world. In New Zealand, the Saanen has also been used to upgrade stock originally introduced by the first settlers, and is the most widely kept of the dairy breeds.

An improved Saanen breed such as the British Saanen is an excellent choice where the priority is for a good dairy animal. The volume of milk produced is often greater than that of other breeds, an important aspect if the milk is to be sold. If it is to be turned into yoghurt, then Saanen breeds may not be the best choice. Yoghurt needs quality milk with a good level of proteins if it is to set properly without the use of artificial setting agents.

The milk of Anglo-Nubians or British Toggenburgs may be more appropriate, but it is important to stress that it is individual 'strains' of goats which are important, not necessarily named breeds. A correct level of feeding will help to ensure that the protein levels of milk are maintained, and this aspect is crucial. Ensuring that adequate records of the yields of parent stock are available when buying stock is also important.

Toggenburg

The **Toggenburg** is another of the

Swiss breeds, originating in Obertoggenburg in the north-west of the country. It is a relatively small goat with a lighter milk yield than that of the Saanen. Its colouring is attractive; a soft brown with white stripes on the face. The coat is often soft and silky with a proportion of long hair.

Introduced into Britain in 1884, with some subsequent importations, it is now a closed breed in the UK. The result of a shortage of new breeding lines means that its milk yields are smaller in Britain than they are for this breed in other countries. Its attractive appearance, friendly disposition and relatively small milk yield makes it ideal for the family which is not primarily concerned with commercial production.

The **British Toggenburg** is considerably bigger, with the colour varying from light brown to deeper shades of brown, but with the characteristic white facial stripes. It has been developed by crossing pure Toggenburgs with British goats. The coat is generally short, without the long silkiness of the original Toggenburg.

The shape of the face tends to be straighter than the more dish-shaped appearance of the Toggenburg. It is Britain's biggest goat and often has a boisterous temperament to match. Large and friendly, it may need stronger confinement to keep it in one place.

Its milk yield is good with a relatively high level of butterfats, making it a good choice for the commercial herd, particularly if yoghurt production is to be an important consideration. The earlier comment about the importance of 'strain' rather than sole reliance on breed, is valid for all types of goat.

In the USA, the Toggenburg has been selectively bred for increased size and production, and it is the most popular breed, surpassing the Saanen. It was originally introduced there from Britain in 1893, although goats of unknown ancestry had been introduced by the British and Spanish colonists.

In Australia and New Zealand the Toggenburg is a popular breed, with many examples of upgraded animals which have been crossed with existing goats.

In the Netherlands, there is an improved version called the **Dutch Toggenburg**, while Germany has the **Thuringian** version of the Toggenburg.

Alpine

Alpine goats were imported into Britain in 1903 and used to grade up local goats. The result is the **British Alpine** which is glossy black with Toggenburg white face stripes and, in some cases, a white belly. It is quite different from the **French Alpine** which is widespread in France, along with Saanens and Poitevens, and is recognized as a breed in the USA. Here it comes in a variety of colours and patterns, including black, white and mid-brown. There is also a **Chamoisée** variety of French Alpine which is similar in colour to the wild chamois.

The **Italian Alpine** is either black, white or grey, although there are pied varieties available. It is not found in

Britain or the USA. The Alpine, as a separate breed, is found in Australia and New Zealand where it is called the British Alpine.

In Britain, the British Alpine's distinctive colouring often makes it the choice of those interested in pet goats, although its milk yield makes it suitable as a milk provider for the family. It has declined in popularity as a commercial animal, mainly because of lack of bloodlines and its smaller yield by comparison with other breeds.

In the USA, the French Alpine is a popular and productive breed for the commercial as well as the hobbyist goatkeeper.

Nubian

Britain has been responsible for developing one of the most popular and distinctive of the goat breeds. The **Anglo-Nubian** was produced by crossing Indian and Sudanese Nubian goats with indigenous British goats at the turn of the century. The result is an attractive and productive animal with Roman nose and long, floppy ears.

The quality of its milk is good, with plenty of butterfats and proteins, although it produces less in quantity than, for example, British Saanens and British Toggenburgs. It is often referred to as 'the Jersey of the goat world' and is a good choice for those interested in producing dairy products such as yoghurt, soft cheese or ice cream.

The Anglo-Nubian is also popular as a family or pet goat and is frequently seen in farm parks and specialist zoos. It is available in a variety of colours, including black, mahogany brown, grey, white and cream. These colours may be single or in mottled patterns.

The Anglo-Nubian has been introduced to many countries, including the USA, Australia and New Zealand, where the prefix 'Anglo' has been dropped in favour of the name **Nubian**.

Golden Guernsey

The little English Channel island of Guernsey has given the world the Guernsey cow as well as the **Golden Guernsey** goat. It really is golden, with silky hair, often long and wavy, and is popular as a small, pet goat for many people.

Its milk yield is fairly small, making it unsuitable as a commercial breed, but quite satisfactory as a family milk provider. It has a considerable advantage over some breeds in its hardiness and high roughage diet preference.

Its distribution is limited, being confined to Guernsey and mainland Britain. However, there are claims that the female of the French Alpine Chamoisée crossed with a Saanen male will produce progeny very similar to Golden Guernseys, so perhaps the origin of the breed lies in France.

Since being introduced to Britain the breed has been developed by crossings, such as with Saanens.

The **English Guernsey** was recognized as a separate breed in 1975. It has been developed by crossing Golden Guernsey females with Saanen or British Saanen males, then crossing

Feral goats in Spain, the wild descendants of previously domesticated goats

Goats are creatures of the heights, as these Greek goats demonstrate on the rock formations of Rhodes

A Golden Guernsey kid

the offspring with Golden Guernsey or English Guernsey sires for three generations. The hope is that the breed will eventually take its place as a commercial animal as well as one for the hobbyist. The use of British Saanen bloodlines should show increased body size and milk yields.

The coat is often paler cream in colour than the original Golden Guernsey, but again, it may be possible to select the golden factor in breeding programmes in order to increase its likelihood.

British

The name **British** goat denotes a cross from any two pedigree breeds or any indigenous goat which is not one of the other recognized breeds. There are some highly productive and hardy goats in this category, including many in commercial herds. There is also an active English Goat Society which promotes the **English** goat. There is a **Welsh** goat which is preserved in a herd for the provision of a regimental mascot for the Welsh Guards.

The **Bagot** goat is a small black and white goat which was originally the herd animal of the Bagot family. It has a history going back at least 600 years, and is found in private zoos and wildlife parks. Feral goats (originally domestic populations that are now wild) such as the **Cheviot** still exist in the mountainous areas of Britain.

Records indicate that there were originally Irish and Scottish as well as Welsh and English goats. Many were crossed and upgraded with Swiss goats, or became feral.

Feral goats in Scotland and other parts of Britain are currently being crossed with imported Cashmere-producing goats to develop fibre-producing goats. It is to be hoped that in the zeal of this drive, the feral population will not be destroyed.

Norwegian goats

A country which kept its original goats relatively free of foreign influence, and yet developed them as good milking animals, is Norway. Their goats (probably similar to the old pre-Swiss breeds of Britain) include the **Telemark**, **Døle**, **Vestland** and **Nordland**. It would be interesting to see what the effect of crossing some of the current breeds in Britain with the Norwegian breeds would be. It could be a way of regaining some of the hardiness which has been lost in certain of our domestic strains, as well as up-grading their Cashmere value.

Spanish goats

The Spanish goats are not found in Britain, but Spain's entry into the European Economic Community has focused new attention on their goats. They are also of interest as being the ancestors of most of South America's goats, as well as those of the USA which were not introduced by British settlers.

The Spanish conquistadors had taken with them the small, brown goats which are still widely distributed in Spain. They include the **Malaguena** of Malaga province and the **Murciana-Granadina** of southern Spain, dairy breeds which, though small, are highly productive.

When I was in southern Spain in 1987, it was interesting to see how each herd was accompanied by a goatherd and a large dog. They accompanied the goats as they wandered the hillsides, keeping watch over them as they browsed the plentiful scrub. Their potential as high yielders on low concentrate, high roughage diets could well be utilized by other countries, including Britain, in future breeding programmes for hill farms and other marginal areas.

It should be noted that the name **Spanish** goats is used in the USA to denote range-kept meat goats as distinct from dairy breeds and Angoras.

La Mancha

This is a relatively new breed developed in the USA and first registered as a breed in the 1950s. It was achieved by crossing local goats of Spanish origin with imported Swiss breeds. Their most noticeable feature is the absence or reduction of the external ear.

I must admit that when I saw them on a visit to America, I did not find it a particularly endearing feature. To me, goats without ears look distinctly odd, but then one of my favourite breeds is the Anglo-Nubian with its long, droopy ears, so perhaps that explains it.

There is no doubt that in the USA they have a popular following and many people keep them as unusual pets. They are not available in Europe, although Spain has a short-eared goat which is probably the origin of this feature in the La Mancha, via goats which were introduced by the Spanish colonists.

African Pygmy

The little African goat is included here because, in recent years, it has become popular in children's zoos and farm parks in Britain and other western countries. It is also the choice of many people who like goats as pets. In Britain and the USA there are societies devoted to these attractive little creatures.

There are basically four types of dwarf goat in Africa, although there are many local variations. These are **West African**, **Southern Sudanese**, **Somali** and **Small East African**. The type which has been most distributed in Europe and the USA is the West African. In these countries it is generally referred to as the **African Pygmy**. It is around 37-45 cm (15-18 in) at the shoulder and is commonly dark brown in colour, although black, white and brown variations and patternings are to be found.

Those who keep them find that a small garden shed with attached dog run is quite adequate. There are

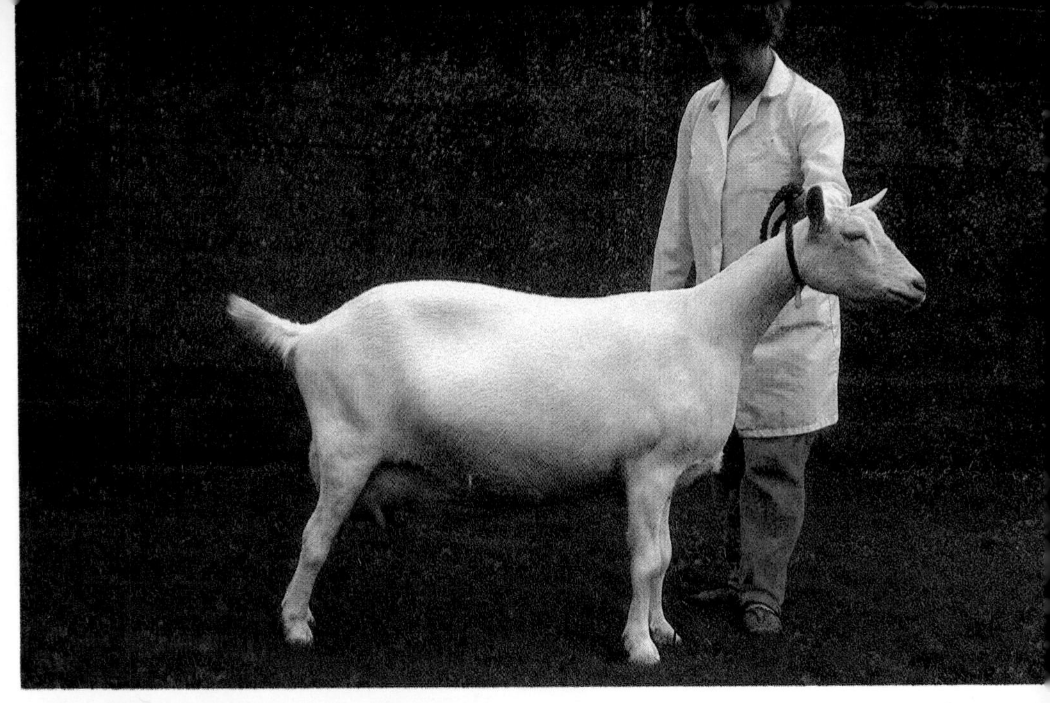

Above: *Pure Saanen milker*

Left: *Swiss Saanens on Alpine pasture in their native country*

Right: *A contented British Saanen views the world*

Cashmere does imported into Britain from Tasmania

several known instances of pygmy goats adapting well to suburban life. They can be trained to use a collar and lead, allowing themselves to be taken for daily walks. The average female will give about a quarter of a litre (half a pint) of milk a day.

Angora

Some of the most expensive goats in the world are pure-bred Angoras. They produce mohair, a fine lustrous fibre much in demand in the quality textile field. It should not be confused with angora wool, which comes from Angora rabbits.

The pure Angora is a most attractive all-white goat with long ringlets of lustrous hair. Originally bred in Turkey, it was introduced into South Africa in 1838. It arrived in the USA in 1848, Australia in 1853 and Britain in 1881, with a subsequent importation in 1885. However, it was not until 1981, when a consignment of New Zealand Angoras was imported, that a rapidly-developing commercial interest got under way in the UK.

Since 1986, the emphasis in Britain has been on the importation of embryo transplants from New Zealand, Australia, Canada and Texas. These are currently being used on Saanen and other British goats.

Cashmere

The name Cashmere was first used by Europeans to describe the goats producing the fine fibre of that name because the trade route for these goats ended in Kashmir. Therefore Kashmir was the place where western traders first saw or heard of the animals, which did not originate there, as some have claimed, but came from the Central Asian mountains, including Tibet, China, Mongolia and Iran.

There is no recognized, fixed breed of goat called the Cashmere, as there is with the Angora. It is essentially a name given to any goats which produce a substantial amount of cashmere fibres. Today, the major producers of cashmere are China, Mongolia, Iran, Pakistan and Afghanistan, with China producing the best quality fibres.

4. HOUSING

Goats need housing which is dry, well ventilated and free of draughts. Purpose-made houses are available, but many people adapt existing buildings. Depending on the scale and number of goats, these may be disused stables, piggeries, garages or other outbuildings.

A garden shed can be adapted for a couple of goats and, in the case of pygmy goats, a large dog kennel and run will suffice. All lend themselves to goat occupation with a little imagination and expertise. The important factor is that the completed house should suit the goats as well as the goatkeeper.

Unless the occupants are the pygmy breed, a minimum height of 3 m (10 ft) is needed, if air is to circulate effectively. The best way of ensuring that it does is to have a combination of ridge ventilation and side inlet. This could be a stable-type door with the top half open, or a hopper-type window which opens inwards from the top. If goats are able to reach this, the glass will either need to be removed,

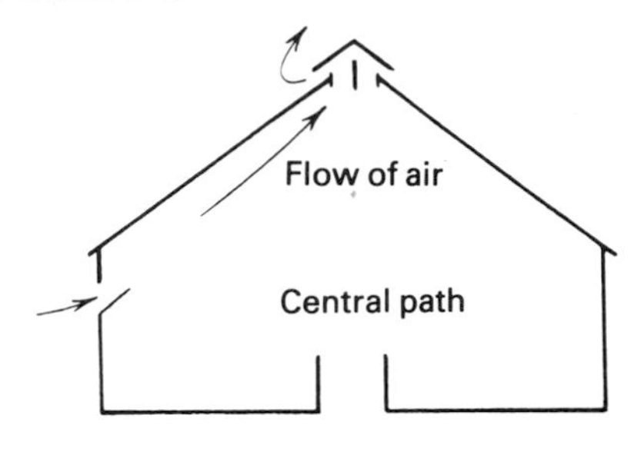

Large communal goathouse seen in section

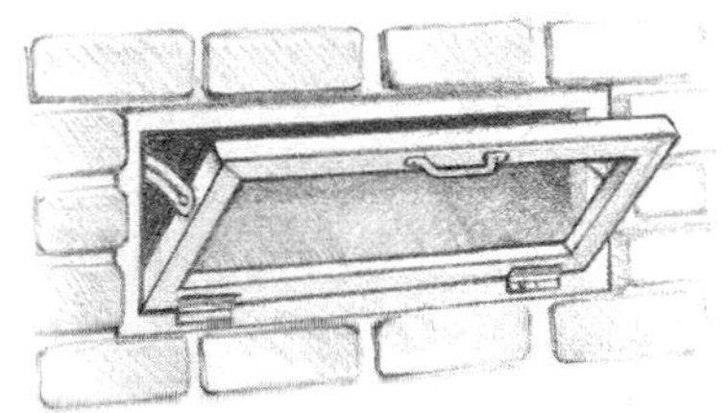

Hopper-type windows give effective ventilation without draughts

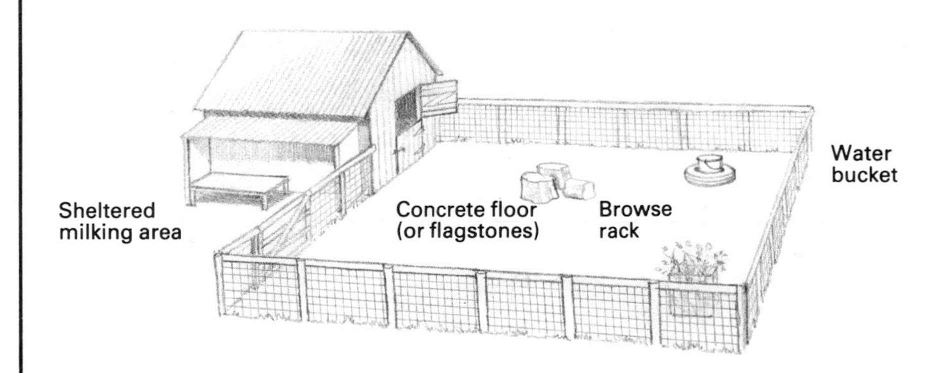

A garden shed adapted to make a house for two family goats

Above: *Swiss Toggenburg type goats*

Far right: *A British Toggenburg male*

Right: *Thuringerwald male – the Thuringian strain of the Toggenburg goat in Germany*

or protected with wire mesh. A small house such as a garden shed will have adequate ventilation from the door and window.

The key factor is that humidity should not exceed 75% and should ideally be around 60–65%. Small humidity meters or hygrometers are available in most garden centres and it is a good idea to put one of these in the goathouse (out of reach of the goats which will otherwise regard it as a tempting morsel).

Walls may be of wood, brick, stone or concrete blocks. All are suitable, but breeze blocks should not be used for external walls because they are porous and let in damp. Wood on its own may prove to be too cold and require insulation to make the house habitable. Insulation panels can be hammered on to the walls but some goats may prove to be destructive of such refinements. A solution may be to use plastic-coated insulation boards, although these are more expensive.

If an outbuilding is being renovated for goat occupation, it goes without saying that a leaking roof is intolerable. It need not be an exhorbitant cost. Heavy duty roofing felt with bitumen is effective in deterring leaks and is widely used in the roofing of small outbuildings. Corrugated iron is strong but has the disadvantage of being extremely cold in winter and far too hot in summer. Modern materials such as **Onduline** are lightweight yet effective, enabling skylights to be inserted which provide light and ventilation.

Electric light is not essential in the goat house, but it is certainly useful, particularly on nightly visits when a goat may be kidding or where there is some other emergency. It is relatively easy and inexpensive to extend an electricity supply in this way, and it more than repays itself in terms of convenience. If the goat enterprise begins in a small way it is always appropriate to look towards future expansion when machine-milking, and the use of other machinery, may become necessary.

The floor of the house is just as important as the roof and walls. An earth-rammed floor is quite unsuitable because rats can burrow in. Bricks, tiles or concrete exclude them effectively. If laying a floor, do ensure that there is a slight slope down towards the door – it makes cleaning so much easier. A larger goathouse will usually have a drainage channel for this purpose.

Layout of a larger-scale goat unit

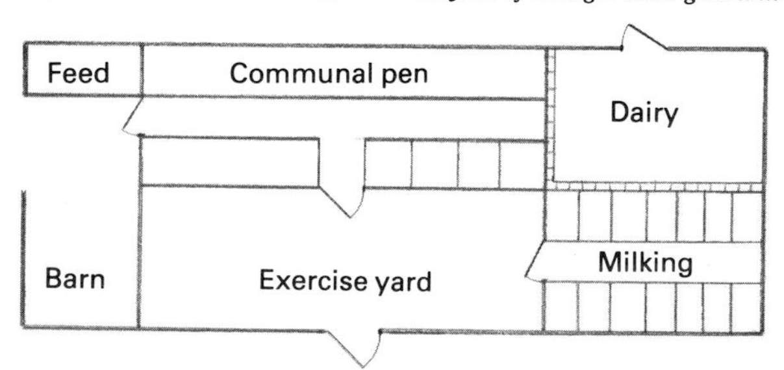

Fibre-producing goats are frequently kept in housing of the Yorkshire boarding type. This is a building which has solid walls up to around 1.2 m (4 ft) in height. The rest of the walls are made of wooden slats arranged vertically, but with ventilation gaps. Polythene housing is also frequently used for fibre goats and as grazing protection for dairy goat herds.

Penning

You need to decide at an early stage whether to have individual pens for the goats or to have communal housing. If only a couple of goats are involved, then obviously the question does not arise; they will share the same house without the need for penning.

The penning system is where the house is divided into a number of individual pens so that each goat has her own quarters. Communal housing is where they are together in one or more large areas: it is generally used in larger enterprises, and a popular system is to have a central path down the middle of the house, with a communal area on either side. This path is ideally wide enough to get a small tractor down it for ease of management.

Plan for a small goathouse for three goats in a converted outbuilding

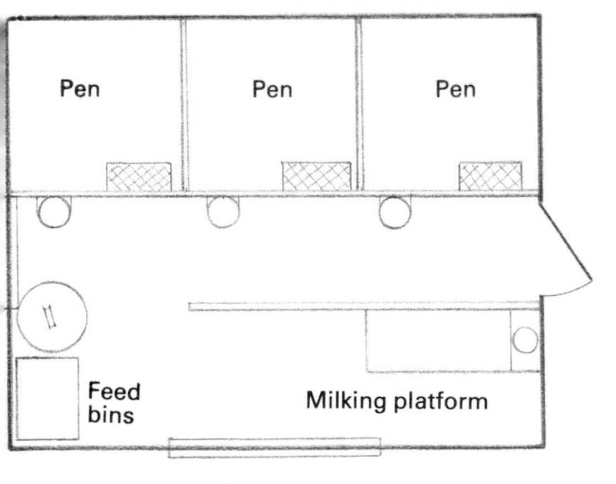

Smaller enterprises tend to favour individual penning. This is more labour-intensive, but with fewer goats labour is not a major factor, and penning does have its advantages. Where a goat is off-colour, for example, it is possible to isolate her where she will not be bullied, until she has had time to recover.

It is possible to buy individual wooden pens, with the purchase including the cost of installation. Alternatively, they are not difficult to construct, but they do need to be strong and well-built. As goats are gregarious herd animals, it is important that they are able to see each other from their pens. One way of achieving this is to use wooden railings covered with **Weldmesh** panels. A warmer alternative is to have solid boarding for the bottom section with mesh panels above. The height of the pen needs to be 1.2 m (4 ft).

Sheep hurdles can also be used to make individual pens. These interlock in any required combination and include gate sections which are fitted as necessary. They are particularly useful if an emergency pen is required. A fully-grown goat requires a pen about 1.8 × 1.2 m (6ft × 4 ft).

A British Toggenburg kid

Swiss Alpine goats

A line-up of British Alpine kids

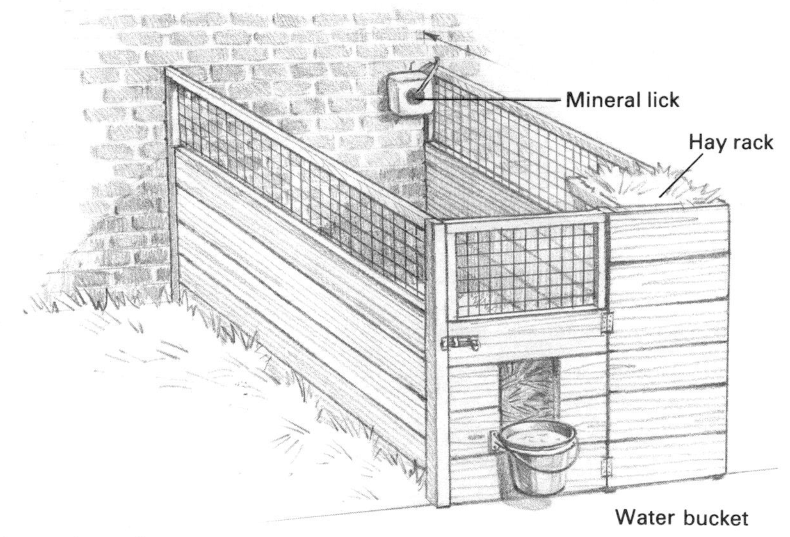

Layout of a pen for one goat

Fixtures and fittings

Individual pens and communal sleeping areas need straw as floor litter. This material is both warm and effective in absorbing droppings, but requires regular cleaning or 'mucking out'. In the intervening times, clean litter is added to the existing layer. The warmth generated in the build-up can be a useful source of heat in the winter months. For this reason, clearing the litter is often delayed in winter in case of a sudden cold snap.

Each pen needs a hayrack, placed at a comfortable height for the goat to help herself. Ones with lids are better because goats are wasteful of hay, pulling out more than they need and then dropping it. They are fastidious creatures (despite their reputation for indiscriminate eating), and will refuse to eat hay once it has been on the ground.

It is more convenient for the goatkeeper if the hayrack can be filled from the outside, without having to open the door and enter the pen.

Hay nets of the sort used by horses are not recommended for goats, which can easily become entangled in them, and cases of strangulation have been known.

Bucket holders attached to the outside of the pen will also allow the feed and water buckets to be filled from the passageway. Some goatkeepers do not have a feed bucket attached to the pen, preferring to make it available in the milking area instead. This ensures that milking proceeds in a calm and contented atmosphere while the goats are busy eating.

A water bucket is essential, however, for water should be available at all times. The hayrack, with its share of hay or green roughage, is an acceptable alternative to the feed bucket as far as the goat is concerned, as long as she has her ration of concentrates elsewhere.

If the buckets and holders are placed outside the pen, it is, of course, essential to have gaps through which the goat can reach them.

Goats need access to salt and mineral licks to ensure that they do not suffer from mineral deficiencies. These are available as blocks which are hung in a convenient position. The animal then licks it whenever she feels the need to do so. The pen is a convenient place to position it, but avoid having too long a piece of twine to suspend it. Again, it is the danger of entanglement which needs to be borne in mind.

Where goats are housed communally they will not need individual hayracks and buckets. As a rough guide, each goat will need 0.5 m (18 in) of rack or trough space.

Large herds usually have an automatic water system with a pipeline unit fed from a central tank. The exit points of the pipeline are either into troughs or nipple units. The animals soon learn to drink from the nipples, and these do save a lot of waste.

On a small scale, an automatic system is not necessary. The time taken to refill the buckets twice a day is not excessive. Problems can arise, however, in winter, when the water in the buckets may freeze, and buckets need far more frequent attention; but this ought not to happen if the goathouse is properly insulated.

The minimum temperature for a goathouse for adult goats should be 6°C (43°F), and the maximum should be 27°C (80°F). Kids require a minimum of 12°C (53°F), but they are normally housed together and so help to keep each other warm. Goats do generate a considerable amount of heat and the commercial herd which is housed communally may have less to fear from cold than the small herd of individually housed domestic goats.

Maximum and minimum thermometers, which show the highest daily temperature as well as the lowest night temperature, are widely available in garden centres. It is a good idea to place one in the goathouse, well out of reach of inquisitive caprine noses.

Hay and feed storage

Hay should be stored in a separate area from the goathouse or pens because it quickly absorbs smells. A large goat farm will usually have a hay barn in which to store enough hay bales for the coming year. On a small scale, a shed, lean-to or any covered area is suitable, as long as it is airy and dry.

Wooden pallets are ideal for the floor, keeping the bales clear of the ground. Air can circulate underneath and there is no danger of damp seeping up from the ground. Wooden pallets (such as those used to transport paper supplies) are quite easy to buy second-hand. Alternatively, use planks laid on bricks.

Concentrate feeds need to be in secure bins, preferably of the metal variety, so that rodents are not able to gain access. They need to have secure lids so that any goat which manages to escape and find its way into the feed

Above: *Butana goats of Sudan, one of the breeds that was crossed with British goats to create the Anglo-Nubian breed*

Left: *Anglo-Nubian kid*

Right: *Anglo-Nubian male*

Haybales should be stacked in dry, airy conditions, and clear of the ground to avoid rising damp

storage area is foiled in its attempts at gluttony. This is important, for an animal which does feed in this way can die as a result of overfeeding.

A convenient way of measuring out the concentrates is to use a metal scoop with a handle. These are widely available from goat equipment suppliers, as are purpose-made feed bins. Alternatives are to use a large aluminium or plastic jug as a scoop, and a metal dustbin with lid for feed storage.

The exercise yard

Not everyone will have an exercise yard for their goats. Where extensive grazing is available, as for example in a field or paddock, they will have sufficient exercise and browsing area. Goats in more confined areas, such as suburban or small acreage conditions, are said to be stall-fed or zero-grazed. In other words, all their food has to be brought to them because there is no natural grazing.

If goats are kept in this manner it is essential that they have an enclosed area where they are able to run freely once out of their stalls. It is here that a scrambling area is useful: arrange logs or piled up paving stones to provide several different levels for 'mountain-climbing'. Watching the goats play 'I'm the king of the castle' can be hilarious fun.

The walls of the yard can be of any material which is strong enough to confine goats. This could be brick, blocks or post and wire netting fencing, and needs to be at least 1.3 m (4 ft 6 in) high. Further details of fencing are on pages 53-7.

The ideal yard surface is concrete. This can be hosed down easily, with the run-off going into a drainage channel and drain. The hard surface, in conjunction with a rocky scrambling area, will help to keep the goats' hooves reasonably worn so that foot trimming is less frequently needed.

The yard will need a water bucket or trough. If it is not possible to attach a bucket to the fencing or wall, a good alternative is to place it in a rubber tyre. This will ensure that it is not knocked over by the gambolling goats. A hayrack is necessary to hold the hay or other roughage, and if this is in a corner of the yard it may be possible to put a partial roof over the rack so that it will not get wet in a sudden shower of rain. This also gives the goats a sheltered area to run to whenever there is a shower.

Many larger goat concerns also have yards, not necessarily because they are short of browsing land, but because it is useful as a general marshalling area. For example, when the goats are to be milked, it may be more convenient to collect them together before leading them as a group into the milking area.

Routine tasks such as foot trimming or worming may also be more conveniently performed in a yard, while fibre goats can be shorn there. In severe weather, outdoor browsing may not be possible and a yard is a good alternative, particularly if one area of it is covered over.

The male goathouse

There is little point in keeping a male unless you are keeping goats on a reasonably large scale, as explained on page 80. If you do keep one he must have his own quarters, which will include a house and an exercise yard.

Housing, penning and fencing must be substantial and well built. A male goat is strong and potentially destructive. Flimsy partitions will soon bite the dust. A suitable arrangement is to have a passageway next to the pen so that food and water can be given, without having to go in. If the pen has a second door opening on to an exercise yard, it is a good idea to have the facility of opening that door without having to go into the yard. When the pen needs cleaning, it is simply a matter of closing the yard door from the pen once he is outside. When the yard needs cleaning, you can shut him in the pen.

It should never be forgotten that an adult male goat is potentially dangerous. Children and anyone who is inexperienced with livestock should not be allowed access to the buck's quarters.

The milking area

The area where goats are milked should be quite separate from the sleeping area. This is basically a question of hygiene, for it is important that the milk should be as free as possible from dust and other contaminants. The floor needs to be easily cleaned, and ideally should be hosed down every day; sealed concrete is a suitable surface in this respect. Walls and roof should also be free of dust and flaking particles, but apart from these necessities, the building itself can be quite simple.

Many goatkeepers find that a milking area in the same building as the sleeping quarters can be an advantage, as long as they are separated by a sturdy partition. It means that on wet days the goats do not need to go outside in order to reach a separate building. Having an exercise yard with a partially covered area will provide similar conditions.

On a small, domestic scale, the milking area may even be outside, as long as the weather is dry. I have often milked goats outside in the sunshine, listening to the birds and enjoying the open air. In wet weather a small lean-to area outside the house is satisfac-

Above: *Golden Guernsey goat*

Left: *Golden Guernsey kid*

Right: *British cross-bred Saanen with triplets. Ideally she would not have horns and would not need to be tethered*

tory, although wintery weather calls for more protected conditions. Milking is an activity which requires cleanliness, whatever the scale, but if the milk is to be sold it is absolutely vital to have a permanent, separate milking area.

A goat is considerably smaller than a cow. The udder is closer to the ground and milking can impose a strain on the back of the milker. A milking stand is a means of alleviating the problem. Individual stands are available commercially, as well as those for several goats. Alternatively, they can be constructed by anyone with a reasonable knowledge of carpentry.

For one or two goats an individual stand is sufficient, allowing one goat to be milked after the other. For larger herds, particularly if you are going to use automatic milking equipment, you are probably best off engaging one of the specialist firms that fit and equip goat milking parlours.

The dairy

The dairy is the area where milk is taken to be processed once it has been extracted from the goats. On a small scale, it will usually be the household kitchen. There is nothing wrong with this if due regard is given to hygiene. On a larger scale, it is necessary to have a separate area where specialized equipment can be housed effectively.

The traditional dairy was once far more widespread than it is now. Most farmhouses and smallholdings had one, usually on the north-facing side of the house for coolness. As prosperity grew and farming became more specialized and less geared to the family, many of the old dairies were turned into extra living accommodation or bathrooms. Modern farms which have subsequently become involved in producing dairy produce have often had to build new dairies or adapt existing outbuildings for the purpose.

The ideal dairy should be cool, light and airy, easily cleaned and with suitable working surfaces. Hot and cold water need to be available, as well as a power supply.

Sealed concrete or tiled floors are the easiest kinds to wash. Hosing is the ideal way of dispersing spilt milk, and for this a drainage channel to an exit drain will be necessary.

It is useful to have a deep sink with a double drainer for the effective cleaning and sterilization of bulky items such as milk pails and churns. Working surfaces which are easily scrubbed are an advantage.

The type of equipment in the dairy will be largely dictated by the type and scale of activity. Milk will need to be filtered and cooled after milking. It may require pasteurization. Some people will concentrate on yoghurt production. Others may prefer to produce cheeses. These activities, and the equipment for each of them, are described on pages 104-114.

5. ACQUIRING GOATS

There are several golden rules to observe about buying goats. They are merely the application of commonsense, but the trouble with commonsense, as has often been observed, is that it is not very common!

Learn about them first This may be a truism, but it is surprising how many people buy goats without knowing anything about them. Reading a book such as this is no substitute for practical experience, but it does provide a basis of factual knowledge which is essential. Joining a local goat club is an excellent way of acquiring experience before buying. Goatkeepers in such societies are usually only too willing to help novices by showing them their livestock and offering a fund of good advice.

You can also go on a course at one of various levels of management, from the basic introductory standard to full-blown commercial training. These are run as local authority evening classes, or they may be organized by agricultural colleges or private individuals or organizations. At such courses it is possible to gain a great deal of experience in activities such as milking, giving worming preparations, foot trimming and so on. A potential buyer is then in a realistic position to take charge of goats.

Establish why you want goats This is not always as obvious as it may seem. There are instances of people who decide to keep goats in the mistaken belief that they will keep the grass down, thus dispensing with the lawnmower. Nothing could be further from the truth. Goats are browsers and highly selective eaters, taking broad-leaved weeds and scrub material in preference to grass. This is not to imply that they do not eat grass. Of course they do, but they will not crop it short the way that sheep will.

It is important to have a clear idea of whether the goats are to be pets, family milk suppliers or commercial livestock. The different categories are not exclusive, for they can be all three, but it does help with overall planning if priorities are established. The type of housing, breed of goat and capital outlay all depend on the answer. Most important of all is the realization that the goats must be looked after every day of the year, including holidays and Christmas day. Holidays and time spent away have to be planned well in advance, making arrangements for adequate care.

Prepare the housing before they arrive I include this at the risk of being accused of stating the obvious. But I have known it happen for goats to arrive at a place where no shelter was available for them. They were bundled unceremoniously into the garage, while the family concerned rushed around, trying to assemble straw bales and bits of corrugated iron into what looked like a parody of the

Above: *Bagot goats*

Left: *A fine example of a female English goat with the male in the background*

Right: *African pygmy goat*

three little pigs' house. I managed to persuade them to buy a second-hand shed from a demolition site before the big bad wolf came to blow it all down.

Avoid buying from a dealer or from a market This is not to imply that all dealers and markets are disreputable. The point is that a novice can make an expensive mistake through lack of experience and judgement. It is much better to buy direct from a breeder who will provide details of the goat's parentage and pedigree.

If the animal is a registered pedigree it will have a registration card with all its details, including the signature of the last registered owner. Without this, it will not be possible to have the transfer of ownership registered to the new owner by (if in Britain) the British Goat Society (other countries have similar schemes operating within their national goat-keeping societies).

It is useful to think of buying a goat in terms of buying a car; the general principles which apply are the same.

Most pedigree goats, and many which are not, are earmarked to facilitate identification.

Buy from a CAE-tested herd CAE stands for Caprine Arthritis Encephalitis, a virus infection for which there is currently no cure (further details are given on pages 148-9).

In Britain, there are two monitoring schemes which allow herd owners to qualify for CAE-free status. The Ministry of Agriculture, Fisheries and Food (MAFF) operates a sheep-goat health scheme which involves regular testing of stock. The British Goat Society (BGS) also conducts tests in order to ascertain which herds may qualify for monitored-herd status. Other countries operate similar schemes.

For prospective buyers, it is simply not worth the risk of buying goats from any other source.

Avoid buying horned goats Horns are dangerous! The gentlest of goats can cause injury without meaning to, by a sudden turn of the head. There have been instances of eye loss and other serious injuries where goats have not been disbudded as kids. Disbudding is the painless burning out of the horn buds at a few days old. Consequently, the horns never grow.

There are some breeds, such as Angoras, which have traditionally been left horned. In my view, there is no justification for not disbudding these as well. As importation of this breed continues, and more Angora crosses are kept for various purposes, it is to be hoped that breeders and recipients will follow BGS advice that all goats which are regularly handled should be disbudded.

Parentage, conformation and condition

The individual buyer will have decided, after careful consideration, which breed or type of goat to purchase. If it is to be a pedigree for possible entering in shows, it needs to measure up to the appropriate set of **standards** for the breed. Each breed has a recognized set of standards in relation to conformation, size, colour and markings.

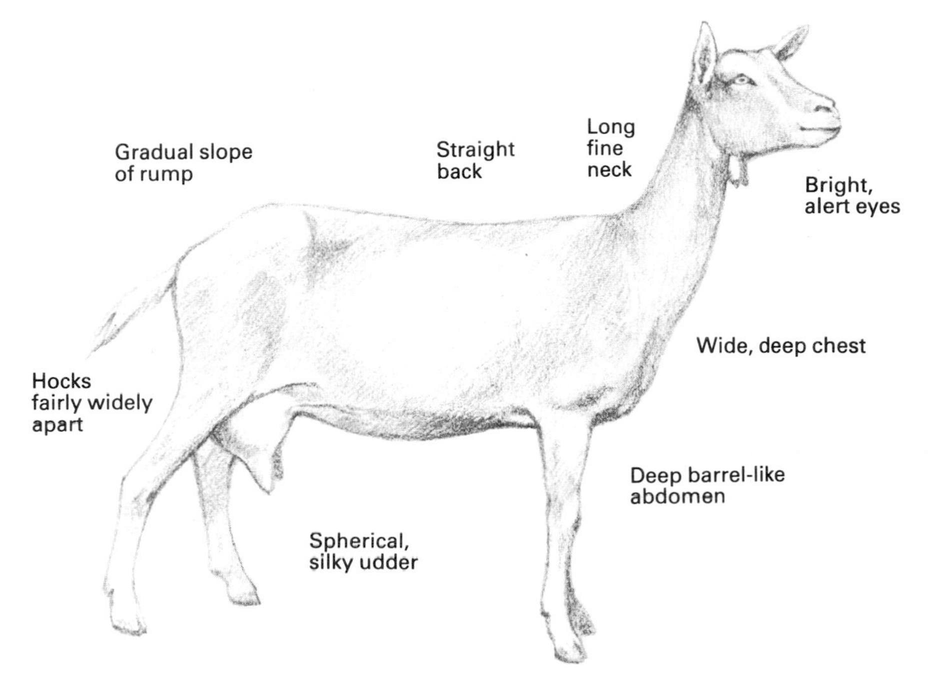

Conformation of a good dairy goat

Most of the popular breeds have a breed society which looks after the interests of the breed and the goat-keepers who are involved with it. The secretary of such a society will provide details of the appropriate standards, and advise on sources of stock.

Parentage In the UK, a dairy goat with the prefix R to her name is one officially milk-recorded by the Milk Marketing Board as having given over 1,000 kg (2,205 lb) of milk with over 3% butterfat in a 365 day lactation.

A stud male whose mother and father's mother have both qualified in this way, will have the symbol § before his name.

Some males also have a dagger prefix, †. This indicates that his mother (dam) and his father's (sire's) dam have both qualified for awards in 24-hour milking competitions at BGS recognized shows. These awards are ★ or Q★ symbols.

Conformation One of the key factors which indicates a good conformation in a dairy goat is the **size of the rumen**. Where this is reasonably large, it shows that the goat has a good capacity for roughage or bulky foods. Rumen size is related to her shape which should be that of a wedge or triangle when viewed from the side, from above, or from the front. Her back, however, should be nice and straight with no marked dip in the middle, a feature which could indicate muscular weakness.

Above: *Angora goats in Australia*

Left: *A good Angora goat should have a generous growth of mohair on the head and neck*

Right: *A herd of Spanish goats on the hillsides of southern Spain*

Legs should be straight, with the back ones fairly wide apart. The pasterns, or area of leg immediately above the back of the feet, should be straight and not turned inwards. The feet should be neat, with no overgrown hooves.

The **head and neck** of a dairy goat are fine and neat. There may or may not be tassels, depending on the breed or strain of the goat. These are long, thin appendages hanging from the area at the back and under the jawline.

The eyes should be bright and alert. Eyes, ears, nostrils and vent should be free of discharge.

Check the **ears** for identification marks. In Britain, the tattoos are in the right ear. Some countries have them in the left ear, while others mark both ears. While checking the head, examine the **mouth** if buying a goat whose age is unclear. The condition of the teeth will give an indication of this.

The **udder** of a milking goat is obviously very important. It should be well attached, indicating that the muscles are in good condition. Any lumps or areas of scar tissue could indicate previous bouts of mastitis and should be investigated. The udder should be smooth and silky to the touch.

The ideal shape is spherical, with the two teats pointing slightly forwards. These should taper towards the end and be of a reasonable size for ease of milking. If the teats are overlarge, kids may experience difficulty in suckling. It should be remembered that the udder and teats will enlarge considerably after kidding.

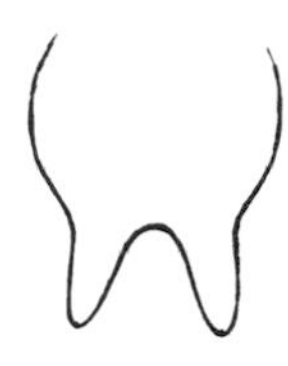

A good, even udder

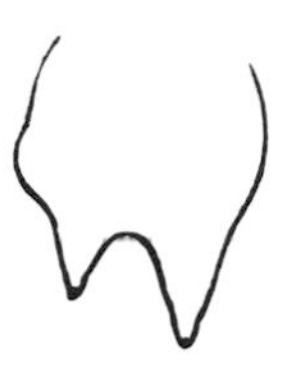

A misshapen udder

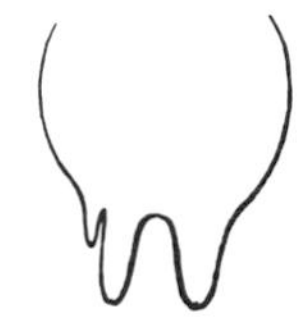

Extra teat: hard to milk

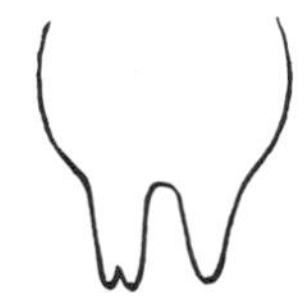

Fishtail teat: hard to milk

Conformation of the udder

Condition Check the coat and skin of the goat to ensure that these are clean and free of external parasites such as lice, mites or ticks. If the coat is coarse and staring, it may be an indication of internal worm infestation. There should be no dry, flaking skin or other sore conditions. It is worth remembering that a goat in full milk production will not be in show condition. Allowances should also be made for those which may have been wintering inside; their coats are bound to have lost condition in that time.

Angoras and crosses The conformation of the Angora goat is quite different from that of the dairy goat. It is much more rectangular in shape, giving a blockier appearance, more like a sheep than a goat. The legs are shorter and sturdier, with heavier muscular growth, making it more suitable as a meat animal than its dairy cousins; and the same is true of cross-breeds between Angoras and other goats (page 115).

Pure Angoras will have long, lustrous curls of fine, white mohair. Angora crosses will have a proportion of rough, straight hair called kemp. It is possible to upgrade crosses by selecting the progeny with relative freedom from kemp for breeding.

Age at which to buy

Overall, you have a choice between kids, goatlings (young females that have not yet been mated), in-kid does (nannies) and milking does. The smaller goatkeeper would not be wise to consider buying a male of any age, unless there is a special reason such as a shortage of good males in the area, or an interest in speciality breeding.

If you buy kids, this has the advantage that they become used to their owner before a possibly difficult period such as milking for the first time is experienced. They do require a lot of looking after, including bottle-feeding if bought young, and you will have to invest a considerable amount of time and expense before you get any return.

Goatlings are extremely high-spirited, but again, there is an opportunity for them to become used to their owner before they take their place in the herd. There is less time to wait before they become productive, and many people buy at this age because the time-consuming tasks such as bottle-feeding and weaning are over.

A five-week-old Saanen kid

Buying an in-kid doe has its advantages, as long as the sire is known and acceptable. She is normally adaptable, without being too excitable, as many goatlings are, but there is obviously less time for you to get to know each other. In recent years, it has become the practice of those interested in fibre-producing goats, to buy in-kid does which have been operated upon to receive embryo transplants of pure Angoras. This is much cheaper than buying pure-bred Angora stock.

Whichever age or type of goat is acquired, it should not be forgotten that the secretary of the local goat society is in an excellent position to help and advise. He or she usually knows who has what goats for sale in the area, and the society frequently has a register of breeders. Such a contact is invaluable.

Transporting goats

Having acquired goats, it is necessary to transport them home. There will of

An adult Golden Guernsey female goat

course be future occasions when you will have to move them from one place to another, such as when travelling to shows or taking a female to a stud male for mating. The most important factors are safety and freedom from stress for both driver and goats.

The ideal form of transport is a trailer attached to a car or other vehicle. Small trailers are available from suppliers and it certainly makes life a lot easier if you have one. It is also invaluable for carrying bulky supplies such as hay bales, straw bales and sacks of feedstuffs. A trailer will have a let-down ramp up which the goats can walk or be led.

Using a normal car may necessitate lifting them in bodily, which is not a good idea for anyone with a potentially weak back; and there is more stress on the goat, too.

If a goat is wearing a collar, a lead can be clipped on to it and she can practise going in and out of the trailer. The best way of doing this is to place food in the trailer, lead her in and let her feed. Then let her out of her own accord so that she does not associate it with any fear. Do this a few times, and start the engine only when she is accustomed to the trailer. Once she has become used to the engine, take her for a short ride before the necessity for a real journey.

However, goats that have just been bought will need to be transported to their new home, so there is no time for such training. They must go in the trailer straight away. Sometimes the vendor will arrange to deliver the goats for you, but it is not a good idea to rely on this. Anyway, there will come a time when you have to have your own transport.

On a small scale, where you have only a couple of goats, it may not be practicable or possible to invest in a trailer. In that case, another vehicle such as a van or even a family car can be pressed into service. An estate car (station wagon in the USA) is better than a saloon because you can put up a partition between the front and the back. Metal grating dog partitions are widely available for this purpose. The

Training to walk up a ramp into a trailer

safety aspect of this is obvious: it is not a pleasant experience to be driving down the highway while a goat is chewing the back of your hair, no matter how friendly its intentions.

A tarpaulin or other waterproof sheet placed in the back will help to protect the vehicle, while straw will absorb any droppings. Most goats are philosophical about travel, although there may be some which have to be lifted in bodily. It certainly helps if she is wearing a collar to which a lead can be attached. Some will sit placidly throughout the journey, particularly if there is hay to munch.

Others are inveterate spectators, intent on standing in order to have a better view of the world outside. The danger here is to the drivers of passing cars who may be distracted by the sight of a hairy caprine face gazing at them. Having a companion in the car who will keep an eye on the goats, and talk to them while you concentrate on the driving, is a great help.

In Britain, it is necessary for goatkeepers to keep a record of movement of their livestock. This is to facilitate tracing the origin of disease in the event of an outbreak. All that is needed is to record the date the movement took place, where from and where to, and the number of goats involved. A sample page is shown below.

This record book may be required for examination at any time. The local police or officials of the Ministry of Agriculture have the right to ask to see it. Goatkeepers in other parts of the world should check whether similar laws are in force in their countries.

Catering for new arrivals

Before going to fetch the new arrivals, the preparations for welcoming them to their new home should be complete. This involves having the house or pens ready with clean straw, a water bucket and a hayrack with hay and perhaps some branches of hazel and willow. If the water is slightly warm, so much the better, for this is something which seems to meet with the approval of most goats. It is better to wait until the following day before feeding concentrates, to allow them to settle down after the possible stress of travelling and being in new surroundings.

Stress is a frequently overlooked cause of problems, but it can affect animals just as much as humans. Paradoxically, stress in humans can often be alleviated if they make an

Movement of livestock record book

Date	No of Goats	Moved from	Moved to
8/7/1987	2	Homestead, Little Place, Essex.	Goat Stud Co., Nearby, Essex

Establishing a good relationship

effort to relax and relate to animals in a sympathetic way. It is no coincidence that nervy, overwrought people tend to have nervy, overwrought goats.

It is essential to establish a good relationship right from the start. Talking to the goats in a gentle voice is not daft. The nature of the conversation, whether it be a review of world economics or the iniquitous price of hay bales, is immaterial. What matters is the **tone of voice**. They will shy away from harsh booming sounds, but respond to gentle, well modulated notes.

I often sing to goats, but as I am Welsh by birth and upbringing this is not surprising. It is still a source of amusement to my children that they, the goats and the chickens were brought up on the same Welsh songs. I have a friend who always plays the radio when she milks the goats. According to her, they like a Mozart violin concerto but do not have the same appreciation for heavy metal rock.

Barbara Woodhouse, the famous dog trainer, once spoke of how she established a relationship with horses by gently breathing into their nostrils through her own nostrils. According to her, it is a way of allowing them to take in your smell and to accept you as someone who poses no threat. I have made the point earlier that goats are sensitive and intelligent creatures. A little effort made to gain their trust and affection will be amply repaid.

6. SYSTEMS OF MANAGEMENT

There are essentially three systems of management when it comes to goats. These are **free-range**, **controlled grazing** and **zero-grazing**. The scale of operations may vary, but the basic principles remain the same, so this information is relevant to all goatkeepers, whether they own a couple of pet goats or whether their herd is a large commercial one.

Free-range This means that the goats are completely unrestricted and are able to select their browsing areas over a wide range of terrain. Most goats are unlikely to have such freedom unless they are feral, or their owners happen to live in extensive moorland or mountainous areas.

Controlled grazing This system is where goats are allowed access to certain carefully-controlled areas of pasture or scrubland on a temporary basis, usually being moved on to new areas in sequence. This prevents overuse of the land and avoids a build-up of parasites.

Zero-grazing This is literally what it says; the goats have no grazing at all, and all food must be brought to them in their stalls or exercise yard.

Most goats are kept under a system which is a mixture of controlled grazing and zero-grazing. In Britain, for example, the cold winters in northerly and easterly regions make it necessary for many goats to be stall-fed for most of the winter. Once the weather improves, they are able to switch to a system of controlled grazing outside.

Some goatkeepers prefer to keep their land to grow hay, not allowing their goats to go on it at all. Instead, the animals are stall-fed and exercised in yards. One of the great advantages of this sytem is that such goats are usually free of internal worms. Their confinement to yards also results in more effective wearing-down of their nails, reducing the need for frequent foot-trimming.

Different people will evolve different systems to suit their own situations, but a typical daily summer routine might be as follows:

- Good morning goats!
- Milkers taken to milking area and fed half their daily concentrate ration while being milked
- Non-milkers fed concentrates in their stalls
- All goats taken to pasture area or to exercise yard (water, browse material and shelter available)
- Regular checks to make sure all is well during the day
- Milkers returned for second milking and given the rest of their concentrate ration
- Non-milkers returned to their pens and given hay and water
- Milkers returned to their pens and given hay and water
- Good night goats!

The grass is always greener on the other side

In winter, or inclement summer weather, this routine will vary, with more browse food and hay to make up for the lack of grazing. I should emphasize that there is nothing sacrosanct about this particular routine. It is but one of many alternatives and goatkeepers will quickly establish what system is the most convenient for them and their goats.

It could be that early milkers may prefer to put their goats back in their stalls until late morning, particularly if there is a heavy and late dew. Others may prefer to give concentrates in a different pattern, or to feed home-grown fodder crops and silage.

What matters is that, whatever the routine, the goats have a balanced diet and the routine is not subject to sudden, drastic changes.

Perimeter fencing

It is no good relying on the traditionally laid stockproof hedge, which is effective for cattle and sheep. Goats will push and eat their way through it. Their mouths can tackle brambles and briars with impunity. If the Prince had been accompanied by a goat when he made his way to Sleeping Beauty's palace in the thorny wood, he would not have needed his hacking sword.

The strongest and most effective livestock fencing is that used for confining deer. It is also the most expensive! This, and the fact that it is much higher than is necessary, makes it hard to justify for goats, unless money is no object.

As a general rule, a 1.3 metre (4 ft 6 in) fence will contain most goats, although the occasional high-spirited goatling may scale it. A satisfactory fence is made by erecting sturdy fence posts, about 3 metres (10 ft) apart, with sheep or pig netting filling the gaps. The posts will need to be braced at the corners for strength. Once a post begins to develop a lean, the

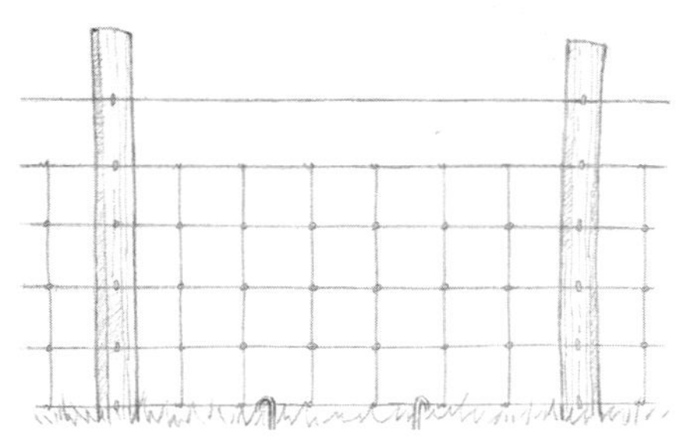

Pig netting is strong but needs an extra strand of wire across the top

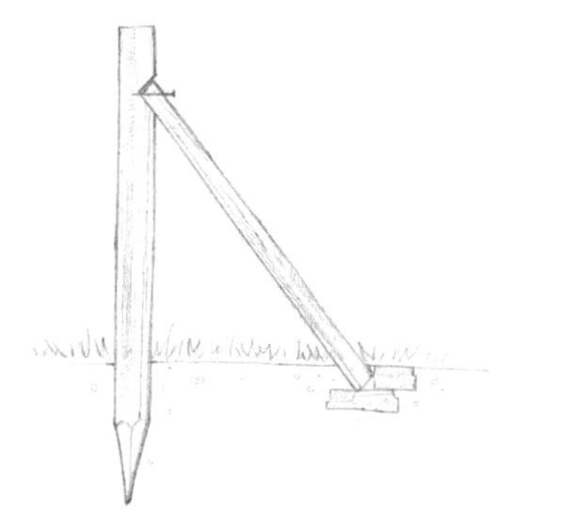

Method of bracing corner posts

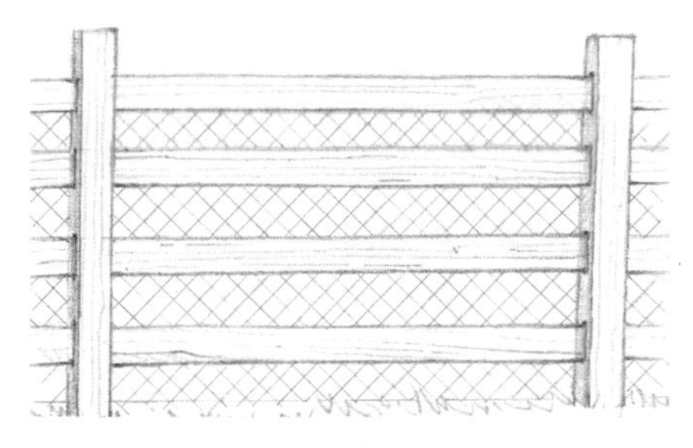

Post and rail fencing should be covered with netting to stop kids and goatlings squeezing through

goats will push against it until it provides a handy stepping stone to freedom.

Sheep and pig netting is not quite high enough, so an extra strand of wire will need to be run above it. This could be electrified, if necessary. Barbed wire should be avoided for it can inflict cruel wounds, and many people would like to see its use banned in the countryside.

Chestnut paling fences are not suitable for goats. They can push their way through them, and the sharp tops of the palings can cause serious damage if a scrambling goat should become impaled (it has happened). If such a fence is already in existence, it can be made safe by covering it with wire netting and folding this over the top. Old poultry netting is suitable for this, although it is not strong enough on its own. It is too weak and sags too easily to confine goats effectively.

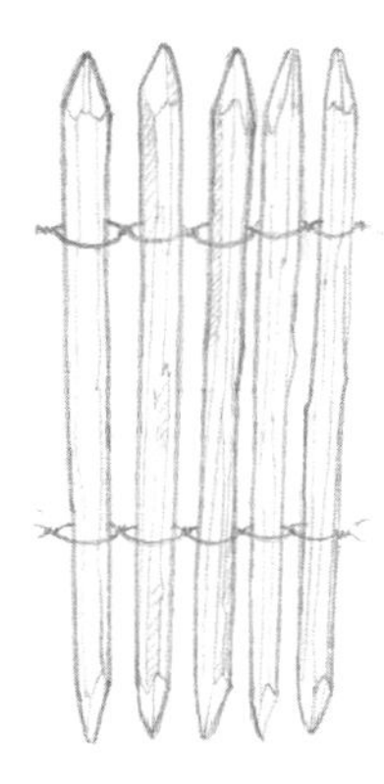

Chestnut paling is dangerous because the goats may become impaled

Post-and-rail fences, of the type used for horses, are suitable, as long as the gaps are filled with netting. If not, kids can squeeze through, while older goats will use the rails as a conveniently-placed ladder.

Traditional stone walls are effective goat confiners, as long as they are fairly smooth-faced. Any jutting stones will remind the goats of the craggy heights of their ancestral homes, and they will use them to launch themselves over the hills and far away.

Electric fencing

Electric fencing can be used in two ways: to supplement perimeter fencing and to control access to pasture on a rotational basis. The first is part of a permanent structure, while the second is a temporarily erected barrier. Before examining them in greater detail, it is appropriate to look at the principle of electric fencing.

The circuit of an electric fence has four parts: (**1**) the controller or energizing unit which is providing the electricity; (**2**) conductor wires which carry the pulses; (**3**) the ground on which the fence is erected; and (**4**) the livestock to be controlled. The controller produces pulses of electricity which output to the fence wires, while an earth lead from the controller is earthed to the ground.

The circuit is not complete until the fence wire is also earthed. This happens when a goat touches it. She receives a shock and moves away. The energizing unit can be mains-operated, or battery-run. Some goat-keepers, living in exposed areas, are able to make use of wind energy by using an aero-generator (a small windmill) to power an electric fence.

Supplementing a perimeter fence If there is an existing perimeter fence or hedge which merely needs supplementing, then possibly one or two strands may be sufficient. We have

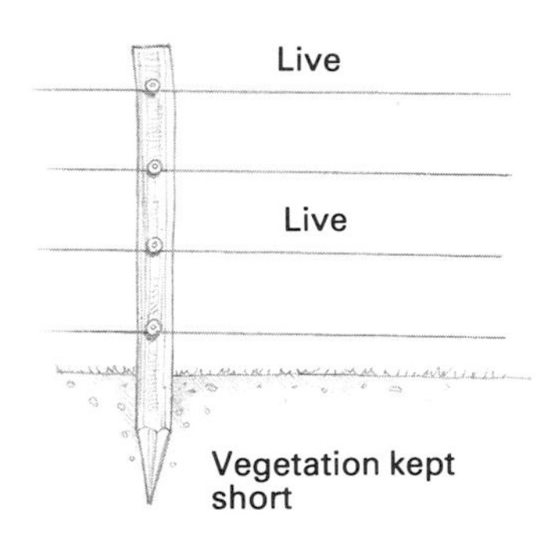

Electric fencing

mentioned sheep or pig netting which requires an extra strand across the top to provide sufficient height, and this strand could well be electrified. Alternatively, a one- or two-strand electric fence could be erected just in front of the existing barrier.

The wires need to be tensioned correctly, otherwise they will tend to sag. It is much easier for two people to erect and tension the wires, rather than for one person to struggle alone. All the equipment and accessories for erecting the fences are available from the manufacturers, and their advice is useful. Some will erect it for you, and there are specialized services that do this as well.

Controlling access to pasture Where access to the land needs to be controlled on a daily or weekly basis, it is not practicable to move a permanent fence. The goats may be strip grazing – eating a section of herbage at a time, before being moved on to the next strip. In such a situation, control needs to be precise.

Lightweight goat electric netting is

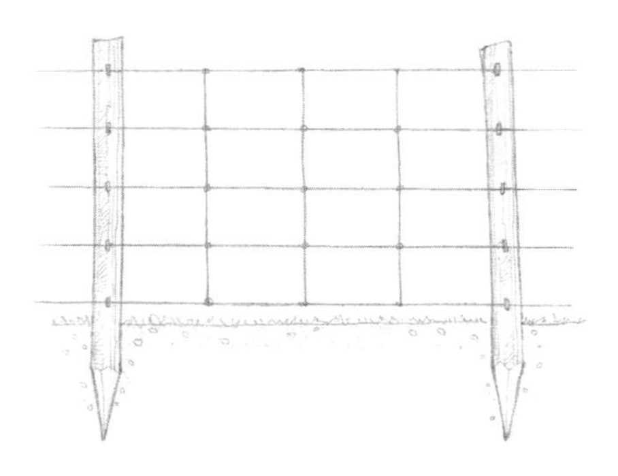

Electric netting – an easily moved control system

available. This can be quickly dismantled, rolled up and re-erected further on. The netting is made of polythene and stainless steel conducting twine and is erected with support poles and ground spikes. It can be used as permanent perimeter fencing as well, but its great attribute, in my view, is its quickness and flexibility of use for different purposes. It is one of the best innovations for goatkeepers for many years.

It should not be supposed that the erection of electric fencing is without its problems. Many things can go wrong and it is important to be aware of the snags and dangers. Any purchased unit should comply with the British Standards Safety Requirements, or its equivalent in other countries. If there is a problem with the unit, the manufacturer or distributor should be consulted. It is not a good idea to dismantle an energizing unit and try to repair it yourself, nor is it sensible to try and convert a battery unit for mains electricity use. The result could be fatal!

It is also important to avoid putting up electric fencing in an area where there is an overhead power supply. At least one fatal accident has occurred, where wire was being tensioned while erecting an electric fence system below. The wire broke, whipped up and touched the overhead power line with tragic results.

Grass must be kept short where an electric fence is erected, otherwise the growth will earth the current and the fence ceases to function. Once the goats have learnt to respect the fence you can switch it off, but you must be sure that they have remembered their lesson.

Training the goats It is hard to generalize about ways of teaching goats because circumstances and individual animals vary so much, but you should initially teach them what the fence can do, rather than simply putting it up, turning the goats out and leaving them to find out the hard way. You will need sensitivity and judgement, for if you merely haul the animal up to the fence you may frighten her badly and possibly harm your relationship with her.

The best approach is to try and lead the animal towards the fence, without pushing her. At the instant she receives a shock shout 'No!' as firmly as possible and then push her away. Immediately make a great fuss of the goat, once she is well away from the fence, stroking and talking to her, perhaps giving her a tit-bit to reassure her.

It is a time-consuming process to do this with every goat, if the herd is large, but it is worth it for the sake of effective and humane management.

There have been cases where goats were confined with electric fence netting and all was well until kids were introduced. The adults had learnt to respect the fence to the extent that it could be switched off. The kids, however, did not know about it. In one incident a kid became tangled up in the netting and was strangled to death. There is really no alternative but to train all the goats and to make regular checks on them while they are confined in this way.

Public footpaths If an electric fence borders a public footpath, there must be a warning notice erected for the benefit of walkers and hikers. Sometimes a footpath may go through the field itself. In this case, it is worth considering a purpose-made gate which is designed to give shock-free access. These are available as free-standing units which can be placed as necessary, and have an automatic closing facility. The current goes through an insulated framework.

Tethering

I dislike the practice of tethering, although there may be occasions when there is no alternative. Tethering involves clipping a chain on to the goat's collar and then attaching the other end to a stake or other fixture in the ground.

The best form of tethering stake is the sort which has a swivel top used in conjunction with a swivel-ended chain. This allows the goat to move in a complete circle without the chain becoming wrapped around the stake. Despite this facility, some goats still

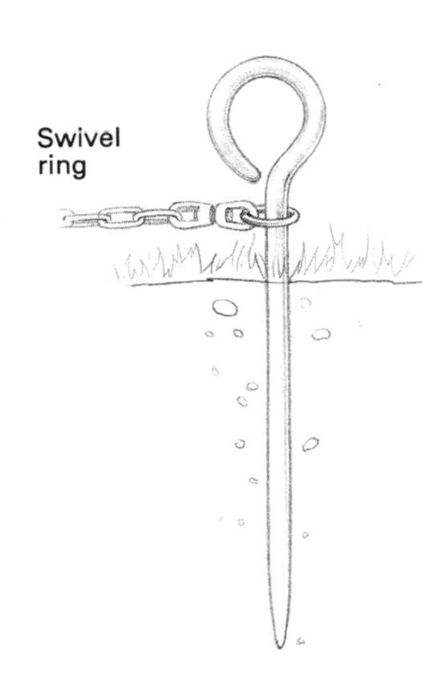

A goat tether

manage to tangle themselves and it is essential to check tethered animals frequently. They are not able to seek shelter in the event of sudden rain or particularly hot sun in these conditions, unless a suitable and transportable shelter is available.

A bucket of water should be placed within reach. This will almost certainly be knocked over unless it is placed in a rubber tyre or other support. A useful one, in this respect, is the metal ring with three prongs which can be pressed into the earth. The bucket is then placed in it.

If two or more goats are tethered, they need to be positioned in such a way that they do not tangle each other's chains. Kids should never be tethered.

Dairy goats usually wear collars. It makes for ease of handling, particularly if they need to be tethered or led. Goat collars are available from goat equipment suppliers. They are available in chrome leather or reinforced nylon and are far better than large dog collars. They are normally available in three sizes, for kids, adult females or adult males.

Alternatives to tethering If there is an alternative to tethering, it is much better to pursue it. If the acreage is small and without sufficient perimeter fencing, it is still possible to use goat electric netting for temporary free grazing.

The other possibility is an exercise yard (see page 36). If you have only a couple of goats it is probably easier to cut the herbage and bring it to them, rather than become involved with all the inconvenience of tethering. Cutting and carrying will also ensure that the greenery is not infected with a parasitic worm burden.

If you are keeping a goat or goats as pets, part of the fun of this relationship is to train them on a lead and take them for walks where they can browse on the way. Teaching a goat to walk on a lead is just like training a dog: she must learn not to pull, and to respond to instructions. With patience, it is possible. I have often taken a goat for a walk up a country lane, stopping when she has spotted a particularly succulent twig. (Pulling on the lead is understandable on such occasions.)

Grazing shelters

Free-ranging goats, as we have said, need shelter to which they can run when rain threatens. A simple three-

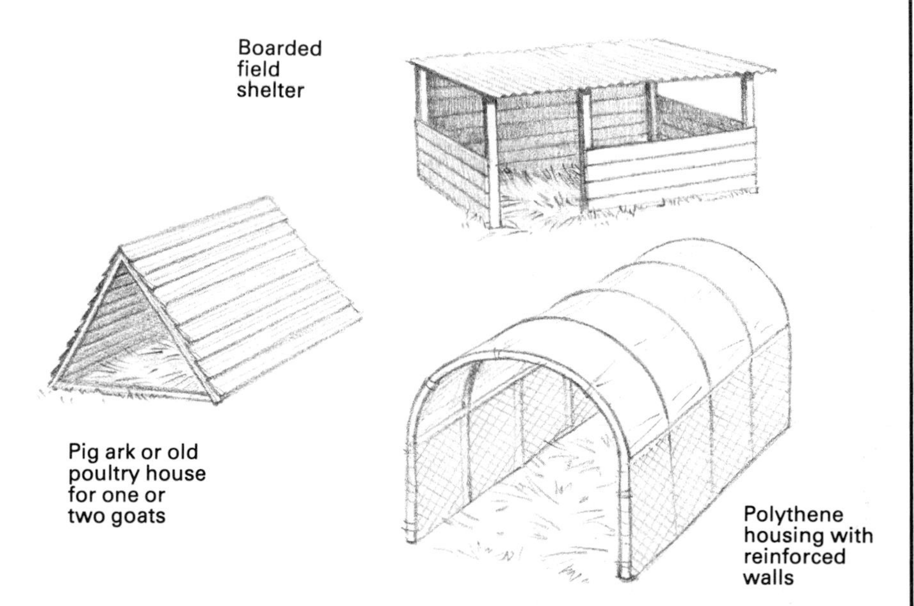

Examples of pasture shelters

sided and open-fronted structure is quite adequate. For a small number of goats, this could be a home-made building, using stout poles and corrugated iron sheets. Disused sheds or poultry houses can be adapted and turned into goat shelters. Any glass should be removed from windows, as well as perches, nest-boxes and so on.

Movable sheep shelters made of plastic are worth considering. These are pegged into the ground and are easily transportable. When not in use, they can be stacked together for storing.

For a larger number of goats, polythene housing makes a suitable grazing shelter. Many polythene housing manufacturers have catered for this need and made livestock shelters in addition to their normal horticultural structures. The bottom half of the wall is normally reinforced with plastic netting or some other suitable material.

An alternative is to use wooden buildings with Yorkshire boarding walls (see page 29).

Pasture management

Pasture is a crop like any other. It is not a playing field or recreation ground. It is a group of plants grown close together which must be fed and maintained if they are to give of their best. There are two kinds of pasture, **permanent** which is never ploughed up and **temporary** where the land is periodically ploughed up and resown with special seed known as a ley mixture. The mixture suitable for goats is not necessarily that which is appropriate for sheep, cattle or horses.

Group of British Alpine goats at pasture

Nutritive value of pasture Goats need a wide range of minerals in their diet. In the feral, or wild, way of life they are able to obtain them by ranging over a considerable expanse of terrain, selectively browsing over a variety of herbage. In more confined, domesticated conditions they have less choice. One of the ways in which the variety can be widened for them is to provide pasture containing a selection of suitable plants.

Grasses are useful sources of protein, carbohydrates and vitamins, but their mineral content is fairly limited. It is the deeper-rooted broad-leaved plants which provide these. The deep roots are able to penetrate to greater depths, bringing up a wider selection of minerals. These plants, from the point of view of goats, are essential reservoirs of the essential elements that they need. A good pasture for goats is therefore one which contains grasses and mineral rich broad-leaved plants.

Sowing a temporary pasture Seed suppliers now sell suitable goat ley mixtures and if you have the use of the land, it is worth considering ploughing and reseeding it with such a mixture. If you do not own the necessary equipment, there are contractors who will come to do it for you.

The area is ploughed over in the autumn and the soil tested to see what the lime and fertilizer requirements are likely to be. You can either do this yourself with a farm-sized soil testing kit, or ask a specialist to do it.

After the land has been ploughed and, if necessary, limed in the autumn, it is harrowed to break down the large particles in the early spring. Then the seed mixture is sown immediately after that. Again, your contractor will perform these operations for you, or you can do them yourself: there are no great mysteries involved, particularly in sowing. This is basically like reseeding a lawn, only on a larger scale. It can be sown in the traditional way, by walking up and down scattering seed, or by using one of the many seeders which are available for sale or hire.

In two to three months, the pasture will be ready for light grazing. As long as grazing is not too heavy, the cropping action of the goats will help to thicken it, the principle being that as plants are grazed down they produce more shoots.

A newly ploughed field is unlikely to need fertilizing in the first season, but it will require some fertilizer for the second one. There is no need to use one of those aggressive chemical fertilizers that disrupts the environment by providing too much nitrogen. A slow-acting and 'kinder' source of nutrients, which is in keeping with the principles of organic husbandry, is calcified seaweed, which is available under various trade names.

It is not a good idea to spread goat manure as a fertilizer for the pasture, for this may introduce parasites to the goats when they subsequently graze.

Pasture should be 'topped' if it gets too long, particularly if the flower heads begin to appear. Doing this will not only maintain the nutrient levels which would otherwise be dispersed by the flowering, but stimulates fresh growth for the goats. If a farm tractor

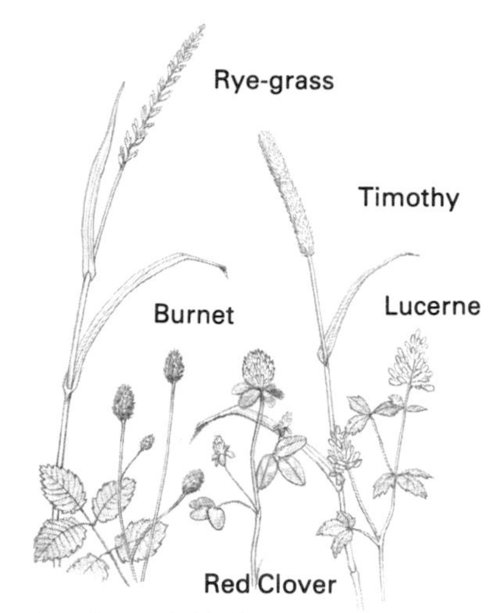

Grazing plants suitable for goats

and cutter is not available you can do it yourself. The ride-on garden tractor, with the blades set at maximum height, is fine.

A suitable ley mixture for goats would include the following:

Perennial ryegrass	Chicory
Timothy	Burnet
Red clover	Sheep's parsley
Wild white clover	Sorrel
Ribgrass plantain	Alsike
Yarrow	Dandelion

This is but one suggestion. Many different mixtures are available, and the seedsman will have his own suggestions to make in the light of specific local soil conditions. Most will make up mixtures to your specification.

Perennial ryegrass is a highly productive grass suitable for a wide range of conditions, while timothy is an adaptable grass which grows more slowly, providing slightly later grazing than the ryegrass.

The red and white clovers are legumes, able to 'fix' nitrogen from the air into their root nodules, thereby increasing the overall nitrogen content of the soil. Alsike is a developed strain of white clover. The remaining plants are all broad-leaved and contain calcium, phosphorus, magnesium and cobalt.

Rotational grazing

The best way to make pasture available is to divide it up so that it is used in rotation. This prevents over-use of the land and prevents a build-up of parasites. The 'resting' area can have other livestock such as cattle to follow, but it is best to avoid sheep for they and goats share many of the same parasites and diseases.

Electric fencing is the most efficient way of controlling access to the pasture (see page 56).

Any pasture which is scheduled for hay cutting should not have goats or any other livestock grazing on it, even for a short time. Once the hay is cut and carried, and the grass has begun to grow again, this is good, clean pasture for them to go out on.

Making your own hay

This is an activity which is possible on any scale. On a field scale it will need a tractor, cutter and baler. If these are not available, a contractor can be hired to do it. It is done in three stages – cutting, turning and baling. If a contractor is hired, the land must be of a reasonable size, otherwise the

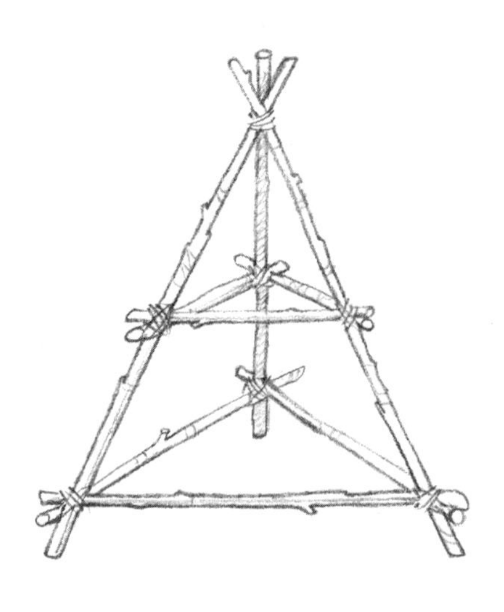

A hay tripod for rapid drying of grass

cost per bale would be greater than if you had bought them.

Cutting The grass is cut at the leafy stage, just before flowering. At this time its nutrient levels are at their peak. The weather is crucial. There is nothing worse than cutting the hay, getting it to the turning stage and suddenly having the rains descending. Britain is particularly at risk, with a climate capable of having rain on any day of the year.

A ride-on garden tractor with the blades set fairly high is perfectly good for cutting hay, and is ideal for the small paddock where farm machinery would be inappropriate and a contractor too expensive. Cutting grass by hand with a scythe is another option, of course, but it is hard work and requires considerable skill.

Drying After cutting, the grass is left until the surface is dry, and then turned over. On a small scale, this is easily done with an ordinary rake. If you put a plaster across the area between the base of the thumb and the index finger on each hand, you will avoid getting blisters.

On a small scale, tripods are commonly used for drying hay. The advantage of this system is that there is plenty of air around it which is conducive to rapid drying. If it does rain, the water runs down rather than soaking through.

Storage Once the hay is completely dry, it can be gathered up into a cart and taken to its storage place. The ideal is to have it baled, but on a small scale this is not possible, unless you are prepared to spend time stuffing it into a container and tying it up.

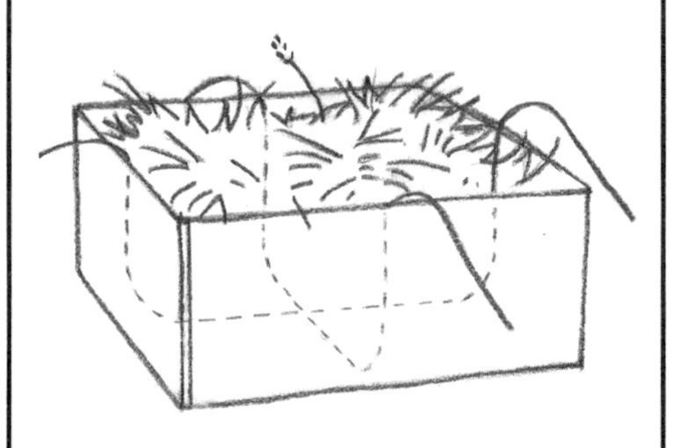

Making hay bales from loose hay

7. FEEDING

The domestic goat requires a balance between **roughage** and **concentrates** in the diet. Roughage is the term for bulk foods such as hay, grass, twiggy growth and herbage generally. Concentrates are foods such as cereals and soya bean meal which have comparatively high nutritional values in relation to their volume.

The goat is a ruminant with a digestive system adapted to dealing with large quantities of bulk foods from browsing. It is essential that a correct balance is maintained between roughage and concentrates. Too much concentrates at the expense of bulk foods will interfere with the workings of the rumen and cause digestive problems.

The **rumen** is just one area of a four-chambered stomach. Bulk foods are chewed quickly and when swallowed go into the **reticulum** and **rumen** to be broken down. This occurs by combination of fermentation and rumination. Rumen bacteria bring about the process of fermentation, while rumination occurs when large pieces of roughage are regurgitated and chewed again, a process referred to as 'chewing the cud'.

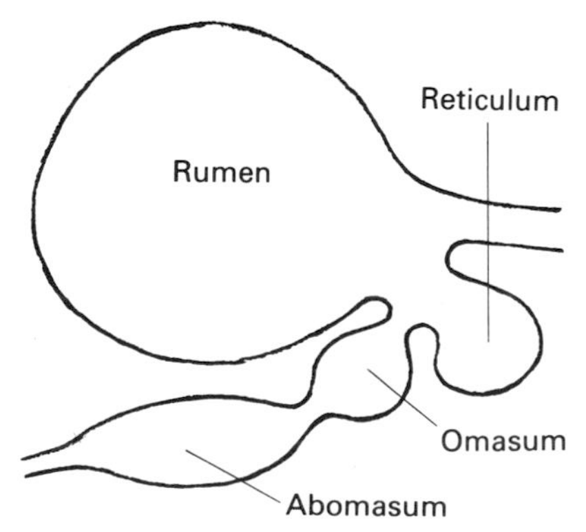

Rumen of the goat

Once the pieces are small enough they are swallowed and passed through the **omasum**, or filtering compartment, into the **abomasum**, or true stomach. Any pieces which fail to get through the filter go back again for another chew.

From the abomasum, the digestive process continues along the normal lines for most mammals. In the small intestine digested nutrients are absorbed into the blood stream through the blood capillaries. The blood then transports them to all the body cells which rely on them for survival and growth.

Any unusable material continues to the large intestine, or **colon**, where water is extracted, leaving the relatively dry **faeces** to be expelled at the **anus**. The surplus water is taken by the blood stream to the kidneys whose filtering action extracts impurities, which are expelled as **urine**.

While the most important single part of the goat's diet is roughage, it is not enough if she is also required to adapt to productive life as a domestic animal. She must also have concentrates, not only to make up for the lack of choice she has in a more restricted environment than that of a feral or wild goat, but to produce milk, meat or fibre.

Maintenance ration:	2 kg (4½ lb) of hay 450 g (1 lb) of concentrates
Production ration for every 1 litre (1.8 pints) milk:	300 g (11 oz) of hay 200 g (7 oz) of concentrates

American readers should note that the British or imperial pint is larger than an American pint. 1 litre is equivalent to 1.8 pints in Britain, but it is marginally over 2 pints in the USA.

It is useful, therefore, to think of the goat's diet in terms of **maintenance** and **production**. A **maintenance ration** is one which provides for the basic needs of the body and its workings. A **production ration** caters for the extras like the production of milk, meat or fibre. As an example, the table above gives a recommended daily ration for an average dairy goat.

The ration formulated above is obviously a generalization because individual goats will vary in their appetite and needs, and who ever heard of an average goat? But it does give a general indication which may be helpful. It is important to remember that the rumen with its roughage always comes first, and the amount of concentrate should not normally exceed 1.8 kg (4 lb) a day, otherwise the balance between roughage and concentrate will be distorted, with possible disturbances to the workings of the rumen.

Hay

As hay is such an important part of the goat's diet, it needs to be the best quality available. It maintains optimum conditions for the rumen bacteria to operate in, and provides the fibrous texture necessary to balance soft, lush growth in the early part of the year. A commonly held belief is that any old hay will do for goats, and the rougher the better. The belief no doubt originated from the idea that because goats eat weeds they must therefore prefer rough grass. It is quite untrue!

Goats eat a wide range of different herbage. They are highly selective eaters. When it comes to a cheap bale of hay and a good quality bale of hay, they will take the quality one. Cheap hay is cheap for good reasons. It was probably cut late in the year so its nutritional value will be relatively low.

The best hay is cut when the pasture is at its most leafy, but just before flowering. The nutritional value at this time is high, and as long as cutting, drying and baling take place quickly, comparatively little of it is lost. A good bale of hay is green rather than yellow, and has a high proportion of leaf in relation to coarse stem. The smell should be that of new-mown hay without any mustiness. The latter indicates dampness and mouldiness which would make it

One of the author's British Saanen goats – Florence – in her pen

not only unpalatable but possibly dangerous.

The presence of seeds in a bale suggests that it was cut after flowering, not before, so the nutritional value is lower than it could be.

Buying cheap hay is an expensive mistake. I once did just that in my early days of goatkeeping. It was the most expensive floor litter I ever had.

The best and most economic way of buying hay is direct from the farm, and in bulk, with sufficient to last for the whole year. Having a trailer is obviously a great advantage here. Considerable storage space is needed, and the ideal is a hay barn which will provide the dry, airy conditions necessary. On a small scale this may not be possible, but it is still worth buying as much as possible early in the season. For storage, turn to page 33.

Some goatkeepers will be in a position to grow, cut and dry their own hay (pages 61, 63). Where land is limited, it is arguable that using it to grow good quality grass or lucerne, rather than letting goats have direct access to it, is preferable. While the goats are exercised in a yard, the land is kept clean and free of worm infestation.

Lucerne (alfalfa) is a popular hay crop, although its use is more common in the USA, Australia and New Zealand than it is in Britain where it is more commonly used in pelleted form. In addition to lucerne pellets, it is available as dried lucerne leaf mixed with molasses.

Both forms are excellent sources of nutrients, but in my view, should be regarded as concentrates rather than roughage. It is easy to forget that pelleted lucerne cannot be as readily dealt with as dried lucerne in the rumen. It is in a much more concentrated form. It is an excellent feed but there have been incidences of goatkeepers overfeeding their goats with them. If you follow the manufacturer's instructions there should be no problems.

Concentrates

Concentrates are what the term implies, foods which are in denser form than roughage, and having a higher nutritional value packed into less space. They include the cereals such as oats, barley, wheat and maize (corn to the Americans), and the pulses such as peas and beans. They are never served whole for they would be too indigestible. They are flaked, chopped, kibbled or crushed and then made up as a concentrate ration. A typical mixture ration, shown by weight, might be:

1 part flaked maize
1 part rolled oats
1 part bran
1 part soya bean meal

If a goat was being fed a basic ration of hay a day, this mixture would provide her concentrate ration. But it could still not be regarded as a complete diet. With the levels of production that are expected of goats, there is an enormous loss of minerals and vitamins taking place, much of it in the milk that is being taken out. In order for the diet to be complete, there needs to be a feed supplement of minerals, trace elements and vitamins. Some of these will be in the existing hay and concentrates, others will come from green food and roots. They can also be added to the concentrate ration as ready-made supplements.

A proprietary coarse goat ration is a mixture of concentrates with minerals, vitamins and trace elements added. A number of feed manufacturers now produce these ready-formulated rations for goats, and they can be regarded as convenience foods in the sense that the average goat-keeper does not need to worry about making up rations.

This has only come about in recent years, as more people have become interested in goats. Before that there were only rations formulated for cattle, sheep, horses and pigs. A request to a feed supplier for a goat ration some years ago brought the suggestion that I get a sack of cattle dairy nuts and one of sheep cubes and mix them up!

Most people will buy concentrate feeds already mixed. Some will prefer to buy the ingredients separately and mix their own. This is usually economic only if you have bulk storage and mixing facilities. It is questionable whether it is any cheaper to mix your own rather than buying ready-mixed concentrates, if the quantities involved are small.

Container and scoop suitable for concentrates

Whatever the situation, it is worth examining the main concentrate feeds:

Wheat Widely available and relatively cheap, wheat must be crushed or clipped before it is utilized in a goat mix. It must not be fed to excess.

Bran This is the fibrous, outer part of wheat and a valuable source of fibre and of Vitamin B. Again an excess can be disruptive because of its laxative properties.

Oats Crushed or rolled oats are preferable because they are easier to digest than whole oats, although they do not store as well.

Barley This is often used in winter rations because of its heating effect, but an excess can cause problems. Use small quantities only.

Maize (Corn in the USA) As flaked maize which has been cooked, rolled and dried, it is in an ideal form for goat mixtures.They love it, often picking out these bits first.

Soya bean A rich source of protein but low in fibre.

Oilseed cake/meal This is the leftover compressed product from the extraction of oils for human use. Rich in protein and energy value, its main drawback is that it does not store for long without going mouldy.

Peas and beans These are high protein foods but need to be coarsely ground or kibbled before use.

Some examples of rations using these feedstuffs, in quantities relative to each other by weight, are indicated below:

1 part rolled oats
1 part flaked maize
1 part bran
1 part oilseed cake

1 part crushed oats
1 part flaked maize
1 part bran
1 part soya bean meal

1 part rolled oats
1 part bran
1 part flaked maize
1 part kibbled peas/beans

These provide basic concentrate rations but need to have added mineral supplements. With additional hay, water, some grazing and browsing as available, goats would be well catered for. Their needs for protein in relation to carbohydrates are in the relation of around 15% to 65%. There is more protein in pasture in spring and summer, so the protein levels can be adjusted accordingly.

Minerals

The precise mineral requirements of goats are still unknown, because insufficient research has been carried out. This in turn is mainly because the goat has not been given serious regard in the past. The situation is changing, thanks mainly to the increasing numbers of people now keeping them, and to their entry into the hitherto 'closed shop' of the farming world. The picture emerging from new research is that the mineral needs of goats have been considerably underestimated.

The goat is much smaller than a cow, yet her yield in relation to her size is much greater. The minerals which are lost in daily milk production have to be replaced. Mention has been made in the section on pasture management (page 61) of the broad-leaved plants such as chicory, ribbed plantain, sorrel, yarrow, burnet and dandelion which can usefully be included with the grasses and clovers of a pasture. There are many other useful sources of minerals in fodder crops and wild foods. Before looking at these it would be appropriate to examine the role of minerals in greater detail.

Calcium and phosphorus These are the main components of the bones and teeth. If not enough is received in the diet, the body draws on its own stores of them. This can have disastrous effects where, for example, a newly-kidded goat produces a flow of milk and there is not enough calcium to cater for it. The result is a collapsed state (see milk fever, page 156).

Deficiencies of calcium and phosphorus can lead to rickets in the young. Calcium is found in legumes, nettles, plantains, weeds such as shepherd's purse, and in dried sugar beet. Phosphorus is found in cereals, bran, oilseed cake and sorrel.

Magnesium Found in the bones and teeth, magnesium is also needed in blood and other tissues of the body. The main sources are grasses, legumes, burnet, chicory, dandelion, yarrow, ribgrass plantain and sorrel. Early spring grasses are sometimes deficient in magnesium until later in the season. The deficiency can give rise to the condition 'grass staggers' (see page 153).

Iodine Essential for the correct workings of the thyroid gland which controls the rate of metabolic activity, iodine is in short supply in the soils of some areas. The use of seaweed based fertilizers on pasture can help to correct the deficiency. Overfeeding of kale and cabbage should be avoided, for they can have the effect of 'locking up' the available iodine.

Cobalt Again, some soils are deficient in this. The element is responsible for the formation of Vitamin B_{12}. A deficiency can produce milk with a tainted taste, and in severe cases, a condition called pine (see page 159). Cobalt 'bullets' can be swallowed, to release the element over a period of several months. Grasses, legumes, yarrow, burnet, chicory, ribgrass plantain and sorrel contain traces of cobalt.

Copper A deficiency of this causes swayback in kids and some areas have a deficiency in the soil. A mineral supplement in the concentrate will provide sufficient levels of this. Too much is toxic.

Sodium In its most common form of salt, this is best made available in the form of a salt lick. This is available either as pure salt or with a mixture of other minerals including cobalt, copper and iodine. Unless you happen to have grazing rights with a beach front, you are unlikely to be able to supply enough sodium to replace the amount which is lost through milk production. A salt lick is essential.

Other trace elements such as zinc for healthy skin, manganese, potassium, iron and sulphur are required. These days, most of them are already supplied in proprietary concentrate rations, although salt is not. In excess, this can have the effect of depressing Vitamin A availability and it is better to make a salt lick available as a separate item. The goats will then lick it as and when they feel the need to do so. The rest of the elements are usually in coarse goat mixes, or they can be added in the form of a supplement such as **Caprivite.**

Vitamins

Goats are unlikely to suffer from vitamin deficiencies if they have pasture, sunlight, hay, concentrates and a good mineral and vitamin supplement in the diet. The availability of fodder crops and wild foods will also increase their choice. The most important of the vitamins together with sources are indicated below:

A	Grass, green foods, carrots, root crops, maize (corn)
B_1	Synthesized in the rumen of the goat; also in grain, peas, beans, bran, green foods
B_2	Green foods, kale
B_6	Green foods
B_{12}	Comfrey

Yeast is a useful source of the vitamin B complex, in addition to the sources listed above. Vitamin B_{12} is found in feedstuffs of animal origin, but comfrey is its only known plant source.

Silage

Silage is produced when grass and other green plants are excluded from air by being compressed in a covered heap. In these anaerobic conditions, bacteria producing lactic acid get to work, and the acid has a 'pickling' effect on the plant material. To produce it properly, it must be done on a fairly large scale. A grass forager is needed as well as a silo to contain the grass.

Although people have tried to make it on a small scale, by using small plastic sacks, for example, the results have not been good. The problem is one of efficient compression and sealing to exclude air. If it is not done properly, there is a risk of circling disease, a condition caused by the soil-borne organism *Listeria monocytogenes* (see page 150).

Any silage which is made can be tested through the Ministry of Agriculture, to ensure that it is safe, but again, this is applicable only to the large scale.

Recently a technique for making silage out of hay bales has been developed. They are automatically shrink-wrapped by machine. Some goatkeepers buy big-bale silage, for in many areas, farmers will deliver such a bale to the goatkeeper's site. In time, it may be possible to have machinery which operates on a smaller scale. Meanwhile, silaged lucerne with molasses is available in fairly small quantities.

The value of silage is as a winter feed, although it must be fed with discretion, ensuring that it comes secondary to hay. Most small goatkeepers are able to manage very well without silage.

Fodder crops

Hay and concentrates are all very well, but from the goat's point of view, it makes a nice change to have a bit of variety, without having to worry about silage. The kitchen garden, or the fodder field is the place to think about! Outer leaves of kitchen vegetables such as cabbage are relished. Carrot peelings, turnip thinnings and pea haulms are greeted with bleats of delight.

On a small scale, this is fine, but the owners of a larger goat enterprise will not think of running down to the goat house with an apple core. They will think in terms of fodder crops grown on a field scale to produce roots for the winter.

The terminology is not important. There is not a lot of difference between growing vegetables in a field and in a kitchen garden, except that the field ones will probably be more at risk from the depradations of rabbits. The following are worth considering.

Kale Sow in the spring on ground which has been well limed in the autumn and manured in the early part of the year. Thin out and keep the soil hoed. If in a field they can be strip-grazed using electric netting to control access.

Garden kale can be cut as needed and put in racks. It is important not to overfeed for two reasons: an excess can bloat the animals, as well as 'lock up' supplies of iodine in the diet, leading to goitre.

Carrots They need a fine seed bed with no fresh manure to make them fork. Carrot fly can be a problem. They can be stored in sand or in a clamp for wintering.

Comfrey Another permanent crop, comfrey is extremely deep-rooted, bringing up huge reserves of min-

Comfrey is a useful fodder crop for goats, and is the only known plant source of vitamin B_{12}

erals. It can be fed green or cut and dried as hay, but beware handling it too much. Some people are allergic to its hairy leaves.

Comfrey is the only known plant source of Vitamin B_{12} and has a reputation for its healing properties in relation to wounds, bruises and skin conditions. Try boiling some leaves and mixing the cooled, strained liquid with vegetable cooking fat. It has been known to clear up nasty wounds and rashes.

Mangolds Sow in the spring in a well manured bed. Do not feed until late winter when they are fully ripened, for they are slightly toxic until then. Avoid giving any to male goats which may suffer urinary blockages as a result.

Lucerne Known as alfalfa in other parts of the world, this is a useful crop for feeding green or cut as hay. It is a legume and will provide nitrogen in the soil. It can be regarded as a permanent crop, with regular cuttings being taken as necessary. It is sown in the spring.

There are many other vegetables and plants which can be grown, but the ones above are the main ones. It is also worth mentioning that the herb garden is a useful source of supplementary food for household goats. It is important not to give too much of anything, for herbs are quite strong and rich in essential oils. They can perhaps be regarded in the way that humans regard herbs in cooking, as natural flavourings and spices to bring interest to meals.

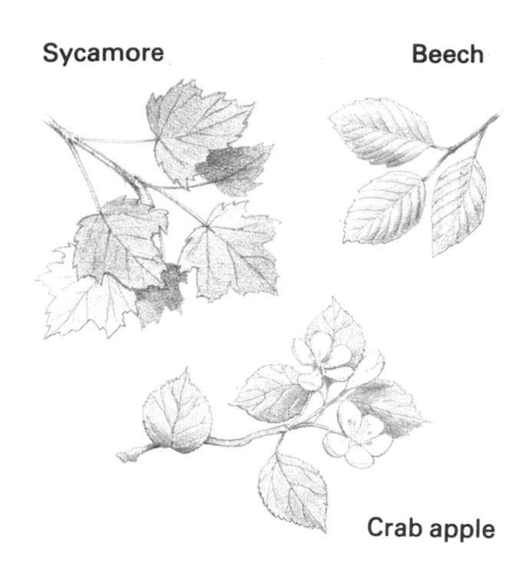

Trees which are suitable browse for goats

Wild foods

There is a whole range of wild foods which can be gathered from the hedgerows, or from the wilder corners of the garden. They are good sources of fibrous, bulky material, minerals and vitamins, as well as being a source of interest and delight to the goats.

Ash	Hazel
Beech	Hornbeam
Birch	Holly
Blackberry	Horse chestnut
Bramble	Lime
Briar rose	Maple
Buckthorn	Plane
Crab apple	Sycamore
Elder	Wild cherry
Elm	Wild plum
Hawthorn	Willow

Most of the common garden weeds are enjoyed by goats and can be added to their green foods when available.

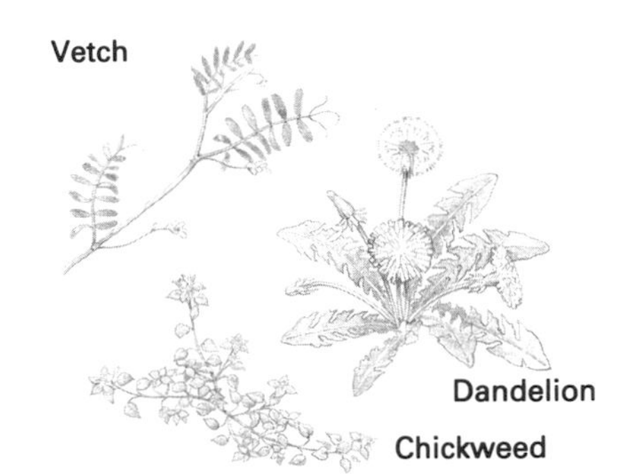

Wild flowers suitable for goats

Some of their particular favourites are:

Young nettles	Knapweed
Thistles	Goosegrass
Clovers	Ivy (not berries)
Docks	Fat hen
Plantains	Vetches
Dandelions	Chickweed

There are also plants which are poisonous and should be avoided. It is difficult to draw a neat line and say that everything listed on one side is good to eat, while those on the other side are poisonous. Life is rarely that simple. It must also be admitted that there are considerable gaps in our knowledge of goats. There has been insufficient research to indicate precisely what plants are toxic to them, and if so, to what degree? This list is therefore confined to the particularly poisonous ones:

Yew	Rhubarb leaves
Rhododendron	Potato haulms
Laburnum	Tomato haulms
Ragwort	Privet
Hemlock	Bracken
Nightshades	Bryony
Foxgloves	Evergreen trees
Autumn crocus	Thorn-apple
Cuckoo pint	

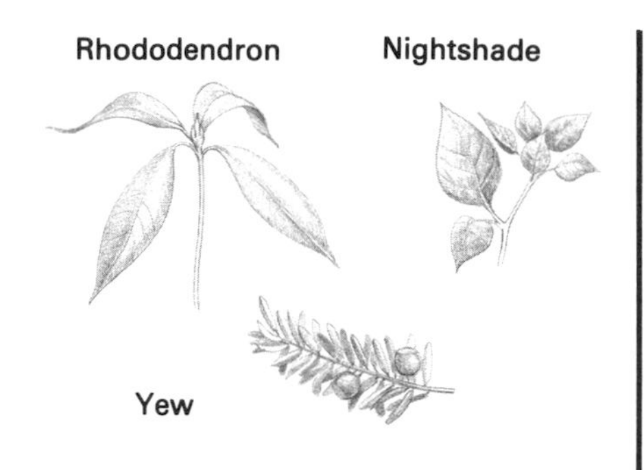

Plants which are poisonous to goats

Water

In South America, goats have been observed to go without water for several days, acquiring their supply from dew and from plants. Domestic and pet goats also occasionally appear to be able to get by without drinking. I once had a goat who never seemed to drink anything from her bucket. I scrubbed the pail until it shone. I tried warm water with a little salt, warm water with a little treacle and warm water with a great deal of pleading. Still she seemed to ignore it.

Then she kidded! before I knew where I was, I was filling and refilling her bucket. It was as if she was making up for the seven years' drought. But it taught me something about goats, and that is, they know what they want, when they want it.

Relative needs for water do fluctuate, depending on bodily needs, amount of succulent foods eaten and general environmental conditions. It

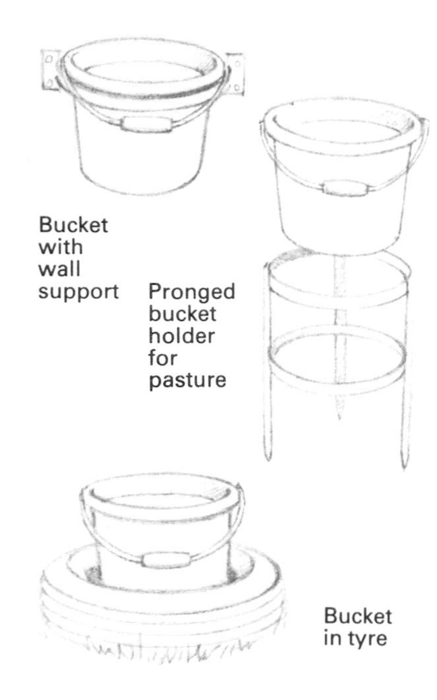

Types of drinking equipment

is important to have water there at all times, because even when little or nothing appears to be drunk, it is probably being sampled when needed. The water should be clean and fresh and changed at least twice a day.

It is particularly important that the male is encouraged to drink if there are any signs of urinary blockage. Males have a tendency to develop stones. If you live in an area of hard water, it is a good idea to give rain water, unless the domestic water supply is fitted with a water softener.

Feeding practice

Like all animals, the goat needs proteins, fats, carbohydrates (starches and sugars), vitamins, minerals and trace elements. These will come from a balanced diet of roughage and concentrates to cater for maintenance and production. Nutritional needs change

Willow is an enjoyable treat for browsers

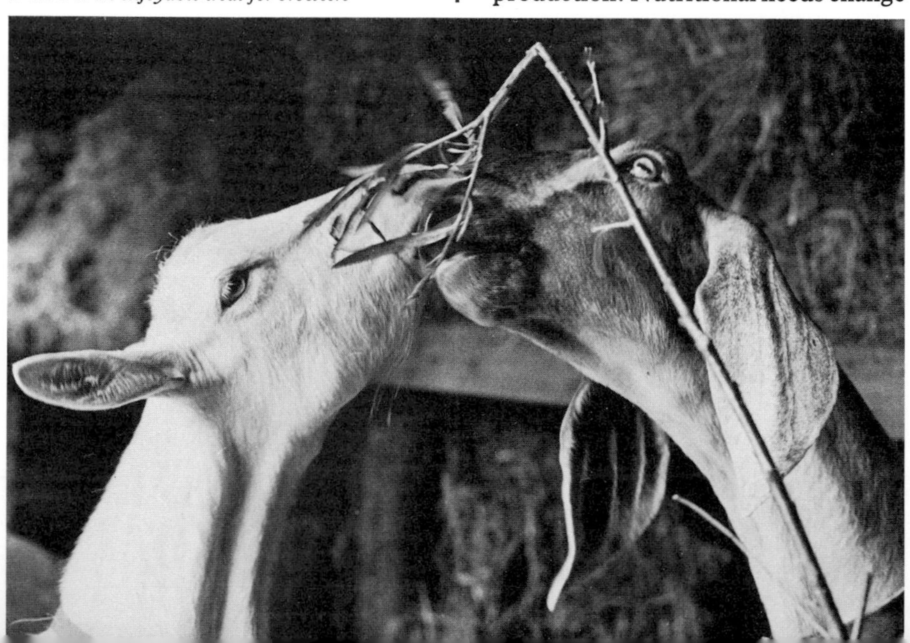

over the year, not only because more of one type of food, such as protein, will be in pasture in the summer, but because the goat itself may have different demands.

A goat drying off during pregnancy will need less protein for a time, but she must then have an increase to cater for the growing kids. A goat kept for its coat will differ from a dairy animal in its requirements. Let us take a closer look at the nutritional needs of specific cases:

The kid It is normal for a dairy goat kid to spend the first four days with its mother. In this way, it receives the essential colostrum, the first thick milk which contains antibodies from the dam. After four days, it will be separated from the mother and put in a communal pen with other kids of similar age. There may be some people who prefer to leave them together longer, but if a milk supply is required, this has to be taken into consideration. It is a personal decision.

The first bottle will not amount to much, but it is important that the kid learns to associate the bottle and teat with food as soon as possible. The various types of bottles and feeders are described on page 90. The following is a typical feeding regime, starting when the kid is five days old:

Start with approximately 0.12 litre (¼ pint) of milk, four times a day, gradually increasing the amount to meet demand, until around 2 litres (approx. 4 pints) a day are taken. By this time, the number of feeds will have been reduced to three a day. Kid replacement feeds are now widely available, to be used instead of the mother's milk. The manufacturer's instructions in relation to mixing and diluting should be followed precisely.

Water to drink, and hay, can be introduced right from the start. The kid will not eat much hay initially, but will play with it and enjoy nibbling. A few concentrates can be given from the second week. As it begins to take more hay and concentrates, the process of weaning can begin from between three to five months. By this time about 225 g (8 oz) of concentrate per day will be consumed, with hay given on an ad-lib basis. A little green food can be introduced as soon as there is evidence of interest. This could start with a handful of cut grass or a few cabbage leaves. A mineral lick should be made available.

By the age of five to six months the kid will be completely weaned and taking ad-lib hay with concentrates between 225-675 g (½-1½ lb) a day, depending on the amount of grazing and other greens or roots.

The goatling By the summer, the goatling will be getting much of its food from grazing and from hay. The concentrate ration will be between 225-450 g (½-1 lb) depending on availability of green foods and fodder crops. Hay should be available to taste. It is important to encourage the development of the rumen as much as possible at this stage. A salt lick and mineral supplement should be made available.

The in-kid doe Once the goatling is in-kid, the concentrate ration should be increased to take into account the developing kids. A gradual increase should be aimed for in order to prevent upsets. By the time she is ready to kid, she will be taking between 1-1.3 kg (2¼-3 lb) of concentrates on a daily basis. A salt lick should be available, as well as a mineral and vitamin supplement in her concentrates. Hay should be fed to appetite, with grazing or other green foods as available.

If it is a mature doe who is in-kid, she will need to be dried off at least two months before kidding. Most will do this naturally. Leaving a little milk behind in the udder each time she is milked will mean that less is produced next time. The concentrate ration can be reduced to coincide with this, remembering the basic rule for a milking doe of 450 g (1 lb) maintenance concentrate plus up to 1.8 kg (4 lb) for every 4.5 litres (1 gallon) of milk produced.

So, if she is giving 1.1 litres (2 pints) a day, her concentrate ration will be 450 g (1 lb) maintenance plus 450 g (1 lb) production ration. This is the lowest it should go to, for giving less than a total of 900 g (2 lb) total concentrate could lead to underfeeding at a crucial stage.

Hay should be made available to appetite, with a mineral and vitamin supplement and a salt lick. As much twiggy growth as possible can be given, but too much lush green foods should be avoided in the drying off period.

Once she has stopped producing milk, it is all systems go again to make sure that the growth of the kids is being catered for. Gradually increase the concentrates to the level it would have been if she had been milking. By the time the kidding date comes this will be around 1.8 kg (4 lb) total daily concentrate with added mineral and vitamin supplement and a salt lick. Green foods can be given in greater quantities again, with as much hay as she will take.

The milking doe The period around kidding is crucial. Give no concentrates at all the day after kidding, but plenty of hay and warm water. A warm bran mash can be given if she will take it (many will not). Concentrates should be increased gradually at first and care taken not to milk her right out for the first few days. These precautions will help to minimize the possibility of milk fever (see page 156).

Once the kids are being bottle-fed, she will gradually be brought to full milk production, following the same formula in relation to maintenance and production that was referred to above. Hay should be available at all times when she is in her stall or in the goat yard. Grazing will be dependent on weather, with green food and fodder crops as available. The greater the range of wild foods and twigs, the better. It is all grist to the extraordinary mill that is a domestic goat.

The male goat If a male is being reared from a kid to adulthood for stud purposes, he will be receiving up to 2.8 litres (5 pints) of milk every day, starting with four feeds and reducing to three. Weaning normally takes place from the age of six months onwards and some goatkeepers continue to give him a little milk right through his first winter. Hay should be available to appetite, with concentrates at around 225 g (8 oz) a day. Grazing and green foods, together with a variety of branches, can be given, but he must have no root crops such as mangolds or dried sugar beet. They can lead to urinary problems.

Once the mating season is under way, the concentrate ration will be gradually increased to a maximum of 1 kg (2½ lb). A salt lick should be placed in his house and his concentrate ration should have a mineral and vitamin supplement.

The meat goat Kids reared for meat are normally surplus males. After four days with the dam in order to have colostrum and a good start, they are switched to a communal pen with other kids. There is no need to separate the sexes at this stage. The males will not normally show sexual behaviour until the age of three months. In order to avoid the need for castration, they can be reared until 10 weeks and then slaughtered.

A milk replacer can be given in place of the mother's milk. Quantities are the same as those detailed for the young stud male above. From 8 weeks of age, a 16% concentrate ration can be given on an ad-lib basis, with hay and water. The concentrate ration used by the Animal and Grassland Research Institute for their trials with Saanen meat kids included barley, soya meal and fish meal. Further details are given in the chapter on meat goats (page 115).

The fibre goat Although the fibre goat is not producing a large volume of milk, she is producing a crop on her back. An Angora goat can grow 2.5 cm (1 in) of mohair a month. If she is not receiving an adequate diet, it will show up initially under a microscopic examination of the fibres, and subsequently to the normal eye. A concentrate ration of around 18-20% protein is required. This can be a normal coarse goat ration, although many of the Angora breeders are using a cube ration because it is less wasteful.

An in-kid Angora will need hay available to appetite, with about 300 g (11 oz) of concentrates. This will increase to around 400 g (14 oz) by kidding time. Once early grazing is available, it can be decreased to 250 g (9 oz) rising to another peak of 400 g (14 oz) in late lactation.

It is normal to leave the kids with the dams, with natural weaning taking place at around three months. Further details are given in the chapter on fibre goats (page 122).

8. BREEDING

Goats breed at definite times of the year, with the onset of **oestrus** or the breeding cycle being controlled by hormonal action. This, in turn, is affected by the decrease in the amount of light in late summer. As a result, the onset of the breeding cycle takes place in autumn. In Europe and the USA goats begin to come into heat from September onwards, while in Australia, New Zealand and South Africa it is from March onwards.

In the tropics, where there is no drastic change in the daily light duration, breeding takes place all year round. As a result, two kiddings a year are common.

The average gestation period is 150 days, although this can vary between 142–162 days, depending on type and size of goat, hereditary factors and on other factors which may relate to the condition of the individual goat.

The reasons why a goat should be put in kid are various, and it is appropriate to look at these in turn:

To ensure a milk supply This is the most common reason, for once a goat has kidded she will produce a regular supply of milk. There are incidences of 'maiden milkers', or of goats lactating without ever having kidded, but the volume is usually fairly low. If a milk supply is the only reason for kidding, then any male goat will do for a 'stud', but if the kids are to be sold, it is better to select a pedigree male.

For future herd replacements It pays to breed the best possible kids as replacements. A registered male with the appropriate characteristics that are required is the best choice. The characteristics will vary, depending on the individual enterprise, but if it is for progeny with a good milking potential, a male with a record of such potential in his family line is obviously the one to choose.

Registered stud males which have a

A British Saanen with three newborn kids several minutes after birth

record of official milk recording schemes and show milking competition results in their ancestry will usually have them indicated by the use of letters, symbols and numbers before or after their names.

The fact that a male does not have these symbols does not necessarily indicate that his line is not a good one; it means that he has not been entered for any competitions. The secretary of the local goat society is in the best position to advise on this question, for he or she is usually well aware of which are the best males in the area.

A high peak yield is not necessarily the most important factor. Duration of yield, particularly over the winter months, is often more important when it comes to ensuring regular milk supplies for customers. For yoghurt production, a quality milk producer with good levels of proteins and butterfats is the most important factor, rather than volume. For this, a good Anglo-Nubian male could be the best choice.

To upgrade fibre producers Those who are interested in mohair, cashmere or cashgora production will wish to choose an Angora or Cashmere feral male for crossing with their females. The progeny can be selectively bred over several generations to improve their potential as fibre producers (see page 125).

To produce kids for meat The best choice for kid meat production is an Angora male. The squarer, blockier conformation produces a relatively quick growth rate with heavier muscle tissue distribution on the upper legs. Crossing between Angora males and feral females to produce fibre goats and meat kids is well established in areas of the world where the Angora has been bred for a comparatively long period. In Britain, it is still at the research stage but is likely to develop in parallel with the fibre-producing industry as a whole.

On a small scale, the most likely source of goat meat is the unwanted male kid. At present, surplus male kids have little value, unless they are Angoras or come from a good line, and most people either have them put down at birth or rear them for a few months until they can be killed for the domestic freezer. Anglo-Nubian cross kids tend to grow at a more rapid rate than the other dairy breeds which are long-legged and tend to have less meat on them.

To produce kids for sale There has to be a demand to make this worthwhile. It is generally true that the best of its kind will always sell. Applying this principle to goats, only the best quality pedigree goats should be aimed for, or those which have good utility qualities. Angora goats and good Angora crosses are easy to sell at present, but the situation could change if there should be a glut following the boom interest in them.

Whatever the reasons are for choosing to go in for goat breeding, it is important to choose breeding animals which are in tip-top condition. Having them blood-tested to ensure freedom from inheritable diseases and from **Caprine Arthritis Encephalitis** is advisable. Many goat studs will now not allow untested goats on to their premises. From the applicant's point of view it is equally desirable that the stud male is known to be blood-tested.

Age at which to breed

Commercial goatkeepers breed their goats in their first year. This means that a kid born at the turn of the year will be mated in the autumn for kidding in the spring. For this, she will need to be well-grown and in good condition.

Smaller goatkeepers, and the owner of the pet goat, who do not have the same commercial pressures, tend to leave their goats until they are 18 months old before they are mated, so that kidding normally takes place around the age of two years old.

It is also the practice of many goatkeepers to confine kiddings to once every two years. If you have two female goats, you may be able to have them kidding alternately, so that as one is declining as a milk producer, the other is ready to take over when she kids. Unlike cows, which produce milk for a specified period of around nine months, a good goat will continue to lactate for two years, or even longer. This allows them to be 'run through' as winter milk suppliers.

Commercial producers tend to kid most of their goats each year, with the kidding period extended over as wide a range as possible. Early kiddings are particularly important to them because the 'staggered' kidding ensures that overall milk supplies do not fluctuate throughout the year.

The male goat

It is unnecessary for the smaller goatkeeper to keep a male goat, unless there is a particular reason for doing so. He must have his own house and exercise yard, and it should never be forgotten that he is potentially dangerous. Even the most experienced goatkeepers never turn their backs on an adult male.

The other aspect which is worth remembering, if you are thinking of keeping a male goat, is that he smells in a most objectionable manner. It is essential to wear an overall kept purely for the purpose before going to the buck house. The smell is particularly pervasive and his house should be placed well away from a dwelling. Close neighbours have real grounds for objection if the smell reaches them.

The choice of buck for mating purposes will depend on what type of kids are wanted. A good dairying strain will require a male with a good record of production in his female ancestry. If you have a registered pedigree goat, then it makes sense to have her mated with a registered pedigree male of the same breed. The kids will have a greater sale value than those which are not pure-bred.

Most local goat societies produce a stud list each year, in the autumn. This gives the name, address and details of the male goats which are available in the area. The list is circulated to their members, and it is an excellent way of choosing a suitable male. The secretary of the society is a valuable source of additional advice.

Artificial insemination

Goatkeepers who live in an area where there is no shortage of stud billies have no problem about having their goats mated. There are areas where the nearest male goat may be more than a day's journey away, so that it is not feasible to rely on someone else's buck. In this case, keeping one's own billy may be appropriate even for small-scale operations. Alternatively, artificial insemination could be considered. Here, the semen of top quality males can be used, without the need to transport goats or to have the feeding costs of keeping a billy.

You will need to have a fairly accurate knowledge of when the female is likely to come into 'heat', or to be receptive to the sperm. The sperm is sold in 'straws'. One of these is introduced into the vagina of the goat using a special applicator. It is not a practice to be undertaken without proper tuition, and anyone who is interested in artificial insemination is recommended to seek out and enrol in one of the inseminator courses that are held in colleges and the like.

The heat period

The breeding period commences in the autumn. Hormonal action triggers off periods of sexual receptivity on the part of the female. This 'heat' period is often referred to as 'being in season', and lasts for anything from a few hours up to three days. It recurs once every 21 days, either until the goat is successfully mated, or the breeding cycle comes to an end in the spring.

The signs which indicate that a goat is in season are persistent tail-wagging, continuous bleating, a slight discharge from the vulva and general restless behaviour, including a tendency to pick at her food. If the

It is unnecessary for the smaller goatkeeper to keep a male goat

goat is to be mated, there is no time to be lost!

Mating

Once the tail-wagging and bleating are really in evidence, the goat should be mated as soon as possible. If the male is on site, this is a straightforward matter of leading her into the buck's exercise yard, making sure that the yard door is securely closed. Then open the door of the buck house to let him out. If she is really on heat, the buck will normally waste no time. It is a good idea to leave them together until a second mating has taken place; this is normally quite soon after the first. It is advisable to have two people to oversee the proceedings, and to control the male.

If it is a question of taking a goat to a male some distance away, it is better to make a provisional booking before the heat period commences. Most stud owners appreciate that it is difficult to be precise about the date, but prefer to have some knowledge of when it is likely to be, rather than having a telephone call out of the blue asking to come now!

Some studs offer boarding facilities, so that the goat can be taken and left for a couple of days, ensuring that mating will take place at the optimum time. Sometimes taking the goat to a place where she can smell the buck is enough to bring her into definite heat.

Once mating has taken place, the owner of the male will issue a certificate to that effect. This is a valuable document if it is a top pedigree male, for it enables the kids to be registered accordingly. If, for any reason, the mating was unsuccessful, and the goat comes into heat again in three weeks time, it is normal practice for the stud owner to have the goat back again for another try, and without charge. Stud fees are comparatively low, although this is not the case where the buck is an Angora.

Embryo transplants

This technique has been developed recently for increasing the numbers of the expensive fibre-producing goats. Nowadays you can buy female goats (of less expensive breeds) which have already been implanted with Angora or Cashmere embryos. Alternatively, your own goats can be operated on to receive embryos, either at a regional centre or on your own premises. There is normally a minimum number of ten accepted for such transplants.

This highly specialized practice is expensive and therefore suited only to the larger enterprises. It is also viewed with apprehension by some goatkeepers who see it as an aspect of intensive farming, incompatible with their views on humane livestock keeping. Whatever view you hold, it is a method which has enabled the Angora breed to become established in Britain at a speed that would have been difficult to achieve with normal importations.

Out of season breeding

It is possible to bring female goats into oestrus earlier than normal by the use of hormones. The advantage of this is that the earlier kiddings ensure that there is a regular milk supply without drastic fluctuations in the winter period. It is a procedure more likely to be used by the commercial producer who relies on regular sales of milk or milk products. The best time to attempt it is a few months before the normal season. In the United Kingdom, this would be from early July onwards.

The technique is to insert a sponge impregnated with the hormone **progesterone** into the vagina of the goat, using an applicator and antiseptic cream to ensure freedom from bacterial infection. This is left in place for about two weeks, and an injection of **gonadotrophin** is given just before the sponge is removed. Removal is simply a matter of pulling on the nylon thread hanging out of the vagina. The onset of heat should follow in one to three days after removal of the sponge.

This method is currently the best one available, but it does have its problems. Failure rate can be as high as 50% and it is expensive, particularly as the sponges are available only in large packs suitable for herds. The sponges may also become displaced; the usual culprit is another goat which sees the dangling thread and is unable to resist the temptation to give it a tug.

A less 'hi-tech' method of bringing about earlier heat is to try and delude the goats into believing that the nights are drawing in. The way to achieve this is to bring them into their house a little earlier each day, and to ensure that no artificial light extends the daylight. This is not as easy as it sounds, for on a bright sunny day it is difficult to maintain the subterfuge that autumn is imminent. It is by no means a foolproof method, but it may be preferred by those who object to the use of hormone treatment.

Pregnancy

If your mated goat does not come back into season three weeks later, the chances are that she is pregnant. If you do detect heat symptoms, take her back to the male. If the second mating also proves unsuccessful, this may be because she is suffering from a condition such as cystic ovaries, which needs veterinary attention.

Goats which are born naturally polled (having no horn buds) may also prove to be infertile. It is not always the case, but if you have a naturally polled goat, it is unwise to mate her with a naturally polled male, for this increases the chances of her having infertile kids.

Occasionally a goat shows all the symptoms of pregnancy, including putting on weight. Then the uterus expels fluid, but no kid. This is called 'cloudburst' or false pregnancy, and the exact cause is unknown, although it is more common in first-year goats which are being allowed to go into their second year before mating. The same goat usually kids normally the following year.

It is possible to detect whether a goat is pregnant by testing a milk sample. In Britain, the Milk Marketing Board will carry out such a test. There are also private laboratories which offer the service.

For details of how to feed a goat during pregnancy see pages 75-6.

General management of the pregnant doe is also important. She should be wormed in the first month and have her feet trimmed before she gets to the heavy stage where damage and stress could result. She should not be expected to go through narrow entrances or be in any stressful situation such as having to cope with noisy dogs. Other measures are to ensure that an **anti-clostridial** injection is administered a few weeks before kidding. The effect of this is to prevent **enterotoxaemia** in the mother, and the immunity is passed to the kids via the bloodstream.

Kidding

It is as well to state at the outset that it is important that no pregnant woman helps at kidding times. Goats (and sheep) in labour can pass on a viral infection which can cause abortion in pregnant women.

The average gestation period for goats is 5 months (150 days) but kidding is possible at any time from 140 days onwards. If the female is normally communally housed, she should be given her own pen with clean, fresh straw. The water bucket here should be placed high enough to prevent the unlikely possibility of a kid being dropped into the water. Although most goats lie down to give birth, a few prefer to stand, so it is possible for such an accident to happen.

The udder will become fuller for anything up to several weeks before, so this is not a reliable indication that kidding is imminent. Occasionally, an udder will become so full that it causes discomfort to the goat. If she is seen to be frequently kicking her back leg, it is probably the reason. As a general rule, the goat should not be milked at this stage, but a little milk can be drawn off to ease the discomfort.

A certain indication that kidding will soon take place is the softening of the rump area, when two definite hollows develop, one on each side of the tail. The goat goes off her food and frequently paws at the straw in her pen. From this time, regular checking is advisable, but it is not uncommon to find that the kids have arrived without warning.

I once had a goat who specialized in producing kids by magic. She invariably had two kids whenever she kidded, always managing to have them licked clean and dry before they were discovered.

Helping at delivery Once a goat begins to strain, the kids are in the process of being pushed down the birth canal. The vulva increases in size and the first sight of the kid will be when the 'water bag' begins to emerge. If this breaks, as it usually does, the two front legs, with a nose resting on top, will be seen. Once the head and shoulders are through, the rest slips out easily and the goat will usually give a loud bellow. If she has been standing up, the umbilical cord will have been stretched sufficiently to break. If lying down, she will break it herself as she licks the kid.

Without interfering too much, check that the nostrils are free of mucus and that the kid is breathing. If a second kid begins to come, it is best to take the first, rub it dry with a towel and place it in some hay in a warm corner of the pen. As long as the goat can see it, she is unlikely to object, particularly if she has to concentrate on the next birth.

A goat can have any number from one to five kids, although two or three is usual. Once they are all born, the afterbirth will follow, although it could be several hours before it comes away. If it shows no likelihood of doing so, do not attempt to pull it, but consult the veterinary surgeon whose telephone number should be close to hand.

A vet's help is also needed if there is obviously a problem in the actual kidding. A goat which has been straining for more than an hour, to no effect, is in urgent need of assistance. Sometimes the cervix does not dilate properly, a condition known as 'ring womb'.

Another potentially serious condition is when the presentation of the kid is such that it is unable to be voided. Sometimes it is simply a matter of pushing the kid back slightly and manipulating it into the correct position with its head resting on the front feet. If it is not possible to achieve this, the vet should be called immediately. It goes without saying that any physical manipulation of this nature requires scrubbed and disinfected hands. The goat will also need antibiotic treatment in case bacterial infection of the womb results.

Post-natal care Remove the afterbirth from the pen and bury it or burn it later. In the wild, a goat will eat the

Presentation of kids for birth

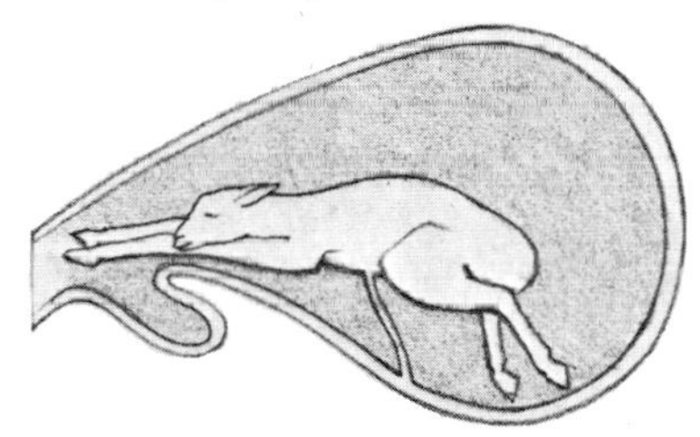

1. Correct presentation for a first kid

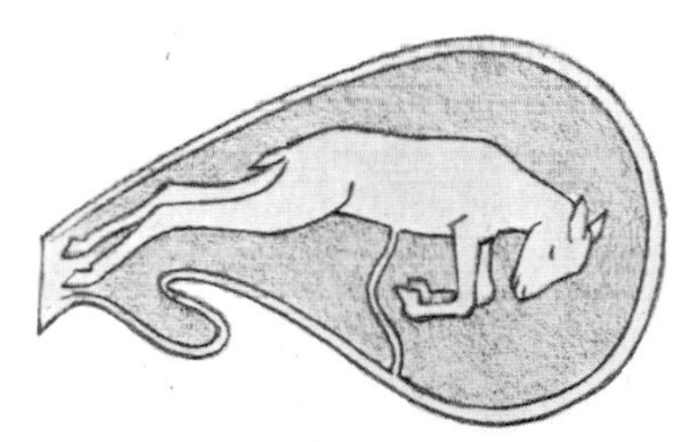

2. Correct presentation for a second kid

3. Back legs need to be manipulated to achieve the presentation in fig 2

4. Head needs to be manipulated to achieve the presentation in fig 1

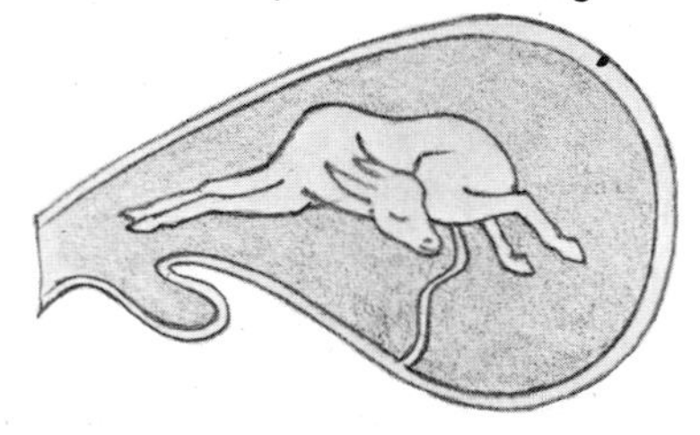

afterbirth, an instinct to conceal evidence that might attract a carnivore, and also a source of much-needed sustenance for an animal temporarily unable to fend for itself. Put in some clean straw to cover the soiled areas, which can be left until the following day to avoid too much disturbance.

The mother will appreciate a drink of warm water more than anything else at this stage. Fill up the hayrack with good quality hay, but do not give any concentrates until the next day. It is important not to give too much food in order to avoid a possible rush of milk formation, leading to milk fever

My experience has always been that goats need warm water, possibly with a little molasses in it, and plenty of hay, but no concentrates at all on the day of kidding. On the next day, my practice was to give no more than 0.2 kg (½ lb) of concentrates, but unlimited hay. From the third day onwards, 0.45 kg (1 lb) would be given, gradually increasing it to keep pace with production levels, up to a maximum of 1.8 kg (4 lb). It is always preferable to underfeed concentrates at this stage rather than to overfeed.

If the mother is heavily soiled she should be sponged clean and then dried. There will be a pinkish discharge from the vagina for a few days after kidding, but this is normal, unless it develops a foul smell. If it does, contact the vet, for it is probably an infection of the womb, requiring an antibiotic. The only other requirement that the mother needs is a worming preparation given within the next few days to ensure that she continues to be free of a worm burden.

The kids

As soon as the kids are born and their breathing has been checked, the umbilical cord should have a spray of **aureomycin** or **tetramycin** to prevent infection. Alternatively, dip the end of the cord in an iodine solution. They should be checked for any defects such as supernumerary (extra) teats, and obvious deformities such as misshapen limbs or defective jaw structures. It is best to put them down immediately, although kids with supernumerary teats could be reared as meat goats.

It is easy to distinguish between males and females. There is no point in keeping the males unless there is a particularly good reason (see page 80). Male kids of dairy breeds have little commercial value in Britain at present, although in countries such as France, there are cooperatives geared to the commercial rearing of kids for meat. Male kids can, of course, be reared for the domestic freezer, and this is the practice of many small goat-keepers.

Occasionally a hermaphrodite kid is born. This is one with characteristics of both sexes. The most common example of this is a kid with an enlarged vulva with a penis-like protuberance. Again, it is best not to keep such a kid. The vet will give the kids a lethal injection if necessary. Alternatively, the huntmaster of the local hounds will arrange for any unwanted kids to be dispatched humanely.

It is important that once the kids are licked or rubbed clean, they should suckle as soon as possible. Most will do this as soon as they are with their mother. Sometimes small kids have difficulty in suckling over-large teats. In this case, it is best to milk out a little milk and put it in a sterilized bottle with a rubber teat. It ensures that the kids have the vital **colostrum**, the first milk which is rich in nutrients and antibodies. Kids which do not have this are unlikely to survive because they do not receive the natural immunity afforded by the antibodies. If there is a surplus of it, it can be stored in the refrigerator or the deep freezer for later use or for emergencies.

Sometimes a premature birth may occur. If the kids are particularly small, they will be unable to feed themselves. Their two main needs are warmth and colostrum. The first will be afforded by a heat lamp such as that used for lambs, or a cardboard box by the kitchen fire. The second must come from the mother by milking out a little colostrum, unless there is some in the freezer for emergencies. Warm it up and add a little glucose, then dribble a little in to the kids' mouths with a dropper.

Small premature kids are unlikely to be able to suck on the normal rubber goat teats. Feeding a little every hour may just pull them through until they are strong enough to suck. It is possible that the mother will reject them if you try to reintroduce them at this stage. If this is so, they will need to be bottle-fed until weaned.

Premature births do not always end happily, but if the effort is successful, it brings a real sense of achievement at having brought back small scraps of life from the brink.

Disbudding

Kids should be disbudded at the age of four days if possible, but certainly before they are a week old. This is the process of burning the horn buds so that the horns do not grow. The procedure is a simple one, but should be carried out by an experienced goat handler or a veterinary surgeon: you will see a demonstration at any goat management course.

The area of the horn buds is shaved back so that it is quite clear where they are. One person will be needed to hold the kid and reassure it, talking gently, while a second person carries out the procedure.

A local anaesthetic is applied while the disbudding iron heats up. Spray-on anaesthetics are normally available from the vet. The iron is a special electrically-powered de-horning iron which is made for goats, and is available from specialist suppliers. It is recommended that one of these is used, rather than trying to make do with something unsuitable and possibly bungling the job. If you have any doubt at all about being able to carry it out adequately, it should be left to the vet or to another experienced handler. No distress or pain should be caused to the kid.

Once the iron is switched on it will heat up fairly quickly. When it is red it is applied to the centre of the bud to a count of six. The edges are then smoothed over with the side of the tool and the procedure repeated with the second bud.

If the horn buds are not completely burnt out, a small piece of horn may subseqently grow. Usually such a

These newly-disbudded kids show no signs of stress and are as eager as ever to play with their young companion

growth reaches a certain size, becomes dislodged and then falls off.

Goats which were not disbudded as kids are occasionally operated on to remove the horns. This is a major operation which can only be carried out under a general anaesthetic by a veterinary surgeon. In Britain, as in many other countries, it is against the law for such an operation to be carried out by anyone who is not a vet.

Earmarking and registration

It makes sense to earmark a kid. The tattoo is a permanent identification mark which can be traced to the owner. In Britain, the practice for those who wish to register their kids with the British Goat Society is to tattoo the inside of the right ear. Other countries have different procedures. In Australia, for example, both ears are tattooed, while some of the South American countries tattoo the under surface of the tail.

It is not necessary to do it yourself. In Britain, most local goat societies have regular earmarking sessions and members take the kids they want registered to these sessions. The kid is tattooed with a code number and the same code is stamped on a certificate. The owner sends this certificate to the BGS, along with details of the kid's parentage and the appropriate fee. The BGS then registers the kid in the name of the owner, recording its code number.

The details of parentage that are required are the certificate of identification of the mother and the mating certificate issued by the owner of the stud male. These are returned after registration is completed. The USA, Australia and New Zealand have similar schemes in operation.

The goat societies are concerned with earmarking goats that are to be registered, but unregistered goats, too, should be earmarked. It is not necessary to have it done through a local goat society. Tattoo earmarkers are widely available from specialist suppliers, but again, the best way to learn the technique is to go and see it demonstrated at a practical course.

Ear tattooing is much preferable to the use of ear tags. The latter are widely used for sheep, as well as for Angora and Cashmere goats. Ear tags can cause damage, bearing in mind the tendency of goats to browse scrub and twiggy growth rather than just grass. The possibility of entanglement and tearing is much greater for ear-tagged goats than it is for sheep. In my view, there is no justification for using ear tags on any goats, whether they be fibre producers or dairy breeds.

Castration

If male kids are kept for any purpose, other than breeding, beyond the age of three months, they should be castrated. Even at this tender age, they are capable of mating with female kids. If kids are being reared together, and the males are not castrated, they must be separated before the age of twelve weeks.

Castration is best performed by a vet when the kid is a few days old. The old technique was the application of a surgical rubber band to the scrotum by means of an elastrator. This stretched the band, enabling it to be slipped over the scrotum. As the tool was withdrawn, the band tightened around the base of the scrotum, restricting the flow of blood to it. This is now considered inhumane and surgical castration by a vet is recommended.

The growth rate of male kids is considerably increased as a result

Earmarking for future identification

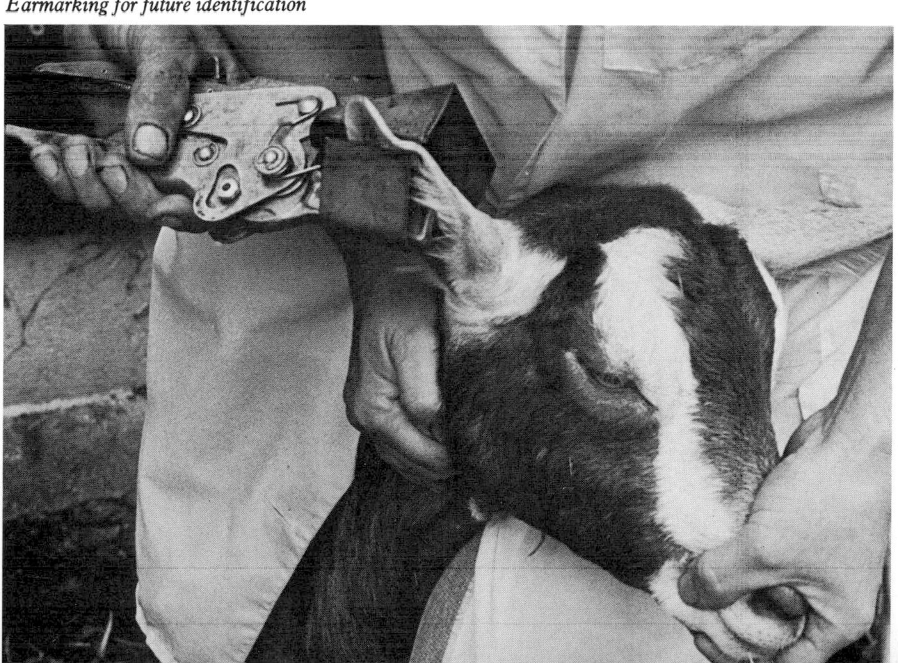

of castration. This is important if the kids are being raised for meat. Another benefit of castration is that the kids can be reared for longer than three months before slaughter without the carcase being tainted by the smell of a sexually active male.

Kid rearing

If goats are kept for milk production, the period for which the kids can be left to suckle is obviously limited, otherwise they are in competition with their owner for the milk. It is normal for kids to be taken from the mother at four days old and bottle-fed. They are best housed together in a large pen with plenty of straw so that they keep warm, as well as providing company for each other. They will soon adapt to these conditions and learn to recognize the bottle instead of the mother's teat as the source of food.

While they are with the mother for the first four days, the kids will take comparatively little milk from her at first, and may go to the same teat each time. This can have the effect of distorting the udder unless care is taken to milk the other teat and balance the amount taken overall.

Bottle feeding This is not difficult but it is time-consuming. It is a task beloved of children and visitors, and a call for volunteers to feed the kids usually has ready takers. Ordinary full-sized lemonade bottles are ideal, for rubber teats made for goat kids will fit them. As the bottles are made of glass they are also easy to wash and sterilize.

The teats may be of the plug-in, pull-on or screw-on variety. There is little difference, although it is a good idea to buy both and see which is preferred. The hole may prove to be too small, particularly as the goats get bigger and become impatient if the milk is slow in coming. Bottles and teats should be sterilized before use, and a proprietary product such as **Milton Sterilizing Fluid** which is sold in chemists' and drug stores for treating baby bottles is ideal.

Feeding a group of kids Where several kids are being reared, giving individual bottles can be very time consuming. An alternative is to use a combined feeder. This is essentially a container with several teat outlets. Warm milk is put in the feeder and the kids feed simultaneously.

A careful watch has to be made to

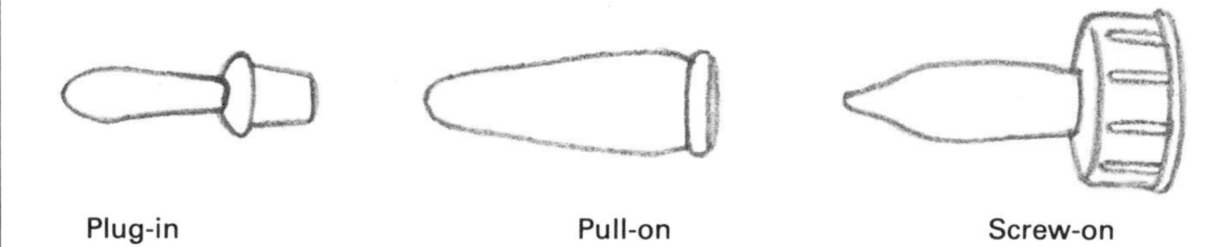

Types of bottle teat

Bottle-feeding kids. Note the angle at which the bottles are held

ensure that they are all receiving equal amounts, for fast drinkers can obtain more than their fair share. It could be that an apparently slow drinker has an awkward teat with a hole that is too small.

Scrupulous attention to cleaning and sterilizing the feeder is necessary. With combined feeding, any infection is likely to affect all the kids.

Milk substitute The mother's milk can continue to be fed in the bottle for a while longer, for it undoubtedly gives the kids a good start in life. As their appetite increases, they can be switched to a proprietary milk replacer feed. These are available as specially formulated kid feeds from animal feed suppliers. It is important to follow the manufacturer's instructions in relation to mixing, and to ensure that no lumps are left. When mixed and warmed, it looks just like ordinary milk.

Warm the milk to blood heat. The traditional way of testing it is the same as that for babies' bottles, by putting a few drops on the inside of the wrist. Hold the kid gently, using the thumb to open its mouth slightly and insert the teat, squeezing it slightly to make some of the milk dribble out. It does not usually take long for the kid to associate the bottle with food, and it sucks vigorously as a result.

At first, it will not take much more than about 0.12 litre (¼ pint) at each feed, but this will gradually increase until it is drinking around 2 litres (4 pints) a day. Initially four feeds a day will be necessary, but by the time the maximum amount is reached, three daily feeds will be sufficient.

It is never too early to introduce hay. In my experience, kids enjoy having a few tufts, even if it is only to play with in the first week. Nevertheless, it accustoms them to it and caters for the instinct to browse. From about the second week, a few twigs of greenery will provide a popular chewing game. At the beginning of the second week, a little concentrate food can be offered. Flaked maize is a good one to start with, or a little of a proprietary goat mix.

Weaning time Weaning, or the discontinuing of bottle feeding, can take place at any time between three and five months of age. By this time, the kids will be eating hay, concentrates and green food, and will have learnt how to drink water from the bucket. There is rarely any need to teach them how to do this. A drinker should be available in their communal pen from an early stage. What tends to happen is that one of them discovers how to use it and the others follow the example. Putting warm water in the drinker is an added incentive to get them started.

At weaning time, the kids should be wormed with a suitable **vermifuge**, or parasitic worm killer. It is also an appropriate time to check their feet in case they need a trim. At about this age, they should have their first **anti-clostridial** injection to protect them against the soil-borne organisms that can cause disease. They should be revaccinated once a year.

Fibre and meat goats Not all kids will be artificially reared. The non-dairying breeds such as the Angora and Cashmere will normally have the kids left to suckle the mothers until they are about three months. They are often housed communally, with the different family groups of mothers and kids together.

When I was in Australia, I was interested to see how these communal groups of Angoras were outside all the time. The climate was mild enough for them to require no shelter other than the shade of eucalyptus trees. They had similar conditions to the feral herds, except that they had the added benefit of food and medication when required.

More information on Angoras is given in the chapter on fibre-producing goats (page 122), while the raising of male kids for meat is detailed in the chapter on meat goats (page 115).

Going out to pasture Depending on

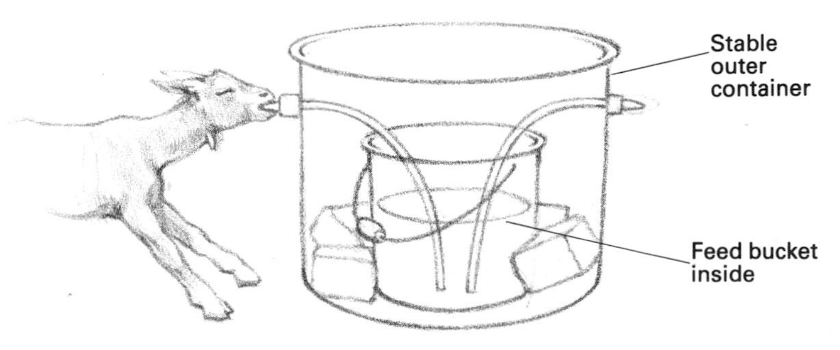

A home-made multifeeder for bottle-reared kids

the weather, kids will be able to go out to pasture when they are weaned. Those that are reared by the mothers will normally accompany them outside earlier, particularly if there is a scrambling area in a yard for them. They will derive enormous enjoyment from playing on a group of logs or boulders.

It is important that they are not subjected to cold, damp weather and that if they go out to pasture, the grass is fresh and clean. Where electric fencing is used, it should be switched on the whole time that the goats are out. (See page 58 about the danger of entanglement.)

Training kids

Kids can be trained from an early age. If they are being bottle fed, they will regard their owner as their mother, so there is already an advantage. One of the ways to start is to get them used to wearing a collar. Kid collars are widely available and a small dog lead can be clipped on.

Practising walking on the lead is an excellent way of getting a kid used to control. A lot of patience is required, with many words of praise where deserved, but also kind firmness if there is any pulling on the lead.

It is important not to let the kids jump up on you. This may be delightful when they are small, but it is a different story when a hefty goatling runs up and throws herself in delight at your chest. A firm 'No!' when they are doing something wrong, but giving them lots of praise and attention when they obey, is well worth it. But, it is too late to start thinking of training them when they are already goatlings.

The correct way to carry a kid

A kid will soon learn to recognize its name, and this in itself is an important aspect of training. For obvious reasons, it is not a good idea to give names to any kids which are scheduled for the deep freezer.

9. MILK PRODUCTION

For many people, the reason for keeping goats is in order to have their own milk supply, particularly if someone in the household has an allergic reaction to cow's milk. The composition of goat's milk is closer to that of human milk than cow's. There is less fat, while the fat particles are smaller, and evenly distributed throughout the milk. These features not only make it easier to digest, but also allow it to be frozen for relatively long periods.

There is a widely-held belief that goat's milk tastes 'goaty'. In fact, if milk is produced hygienically from well-fed dairy goats, and the milk is cooled straight after milking, it should be impossible for most people to distinguish between it and cow's milk. The problem has undoubtedly arisen because, in the past, people did not pasteurize goat's milk, nor did they have the refrigeration equipment that they have today.

Any milk goes off after the first day in such conditions. How many people would be prepared to drink unpasteurized cow's milk that was several days old? It would be equally undrinkable. It is true that goat's milk is slightly more acidic and thinner than

Hand milking the domestic goat. If this goat were standing on a milking platform there would be less strain on the milker's back

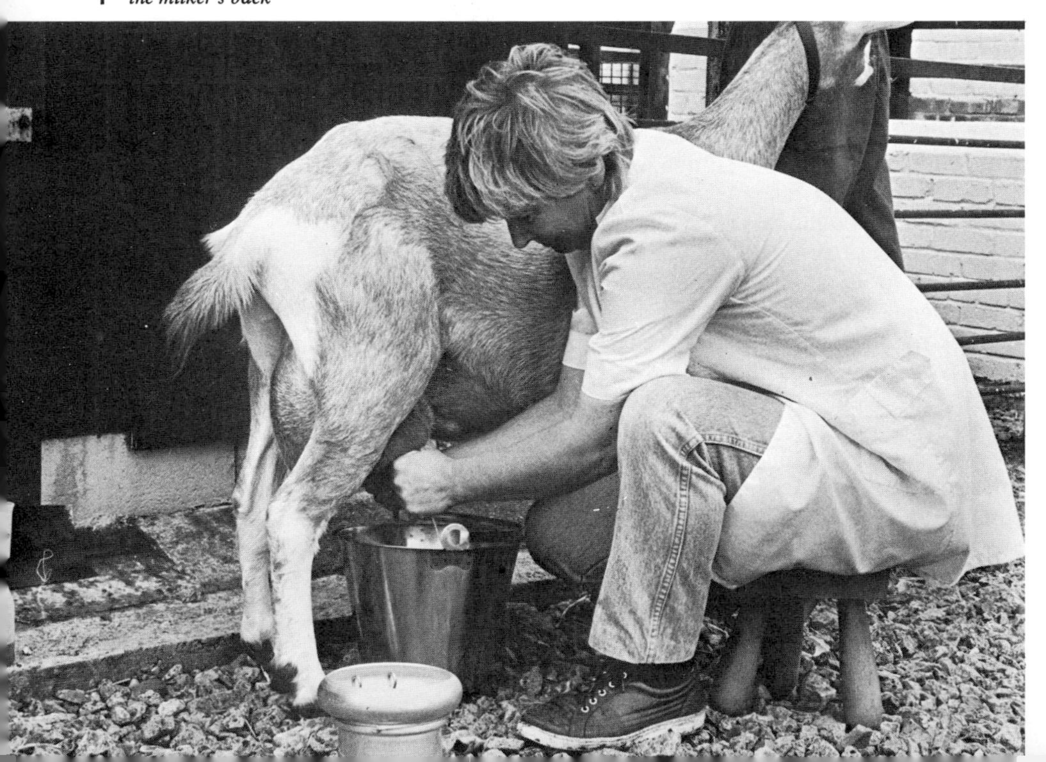

that of cows (the relative pH values are 6.4 for goats and 6.7 for cows), but this is not enough to cause any great difference in taste.

Any dairy animal's output is influenced by what it eats. There are incidences of milk being tainted by certain strong flavours such as cabbage, wild garlic, wild onion and turnips, but these tend to be where an excess is eaten. For example, if a goat is browsing in pasture where wild garlic is plentiful, this may affect the milk.

If, despite scrupulous attention to milk hygiene and diet, a particular goat does appear to be producing tainted milk, there are four possibilities to look into:

(**1**) When milking, have you been splashing the milk into the bucket? This increases the oxygen content and breaks up the fat particles, factors which are conducive to a more rapid production of fatty acids in the milk.

(**2**) Check that the goat does not have active musk glands just behind the horn area. These will give a goaty smell around the head. A veterinary surgeon will cauterize them.

(**3**) Have the goat's urine tested for a possible high level of ketones. These are by-products of the conversion of stored body fats, a situation which occurs when a goat is under-fed. Immediate treatment is afforded by proper feeding and giving glucose as a short-term expedient.

(**4**) Ensure that the goat is not carrying an intolerable worm burden, by giving her a proprietary worming preparation available from the vet.

If all these points have been taken care of but there is still a taint, the culprit is likely to be inadequate milk handling technique. Filter, pasteurize and cool the milk immediately. Then use it as fresh as possible, keeping it in refrigerated conditions in the meantime. I have never encountered a problem of tainted milk which did not disappear when all these requirements were properly heeded, and the most frequent cause was that the milk had not been filtered, pasteurized and cooled.

The udder

A goat's udder is divided into two quarters, each quarter ending in a teat outlet. 'Two quarters' is a paradox in terms, but it stems from the fact that a cow's udder has four quarters and the terminology has been extended to goats, despite the fact that they have only two teats to the cow's four.

Each quarter is separate from the other and is held in place by suspensory ligaments. Milk constituents are delivered by the blood stream into the secretory tissue of the udder (the **alveoli**). From here, milk drains into the central udder cistern above the teat.

A section of tissue temporarily closes off the udder cistern from the teat canal when the top of the teat is squeezed during milking. Squeezing the lower part of the teat thus propels milk out of the teat opening. Once released, the canal fills up again from the cistern.

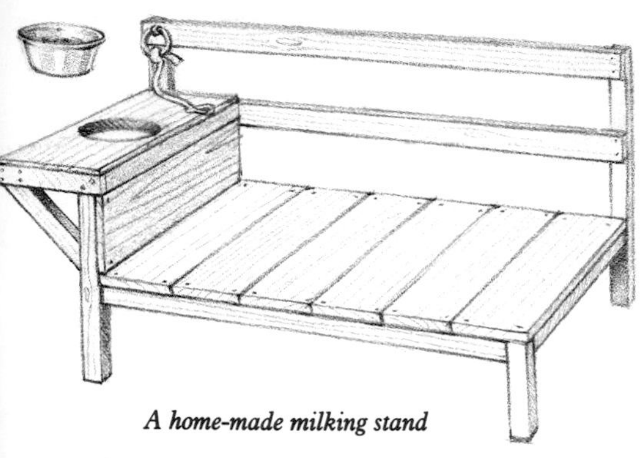

A home-made milking stand

MILKING As I have said in the chapter on housing (page 25) it is very important to have a milking area separate from the sleeping quarters. On a small scale, this need only be an area that is easy to clean, probably with a single milking stand. On a larger scale, particularly where milk is to be sold, a purpose-built milking parlour with easily hosed down surfaces is recommended. The area should ideally have hot and cold water, and hand washing facilities.

Machine milking There are basically two types of milking machine system available:

(1) A portable, self-contained and wheeled unit suitable for one or two goats at a time. This is electric or petrol driven and the milk is pumped into a specialized 'bucket' which filters the milk as it enters.

(2) For larger numbers, a pipeline system is normally used. This has the individual teat clusters leading into a central pipeline which pumps the milk into a bulk tank. Here the milk is filtered as it enters, and is immediately cooled to 5°C (41°F).

Machine milking of goats. The animals are yoked into position on the platform and provided with food to keep them occupied. Their udders are then washed prior to milking

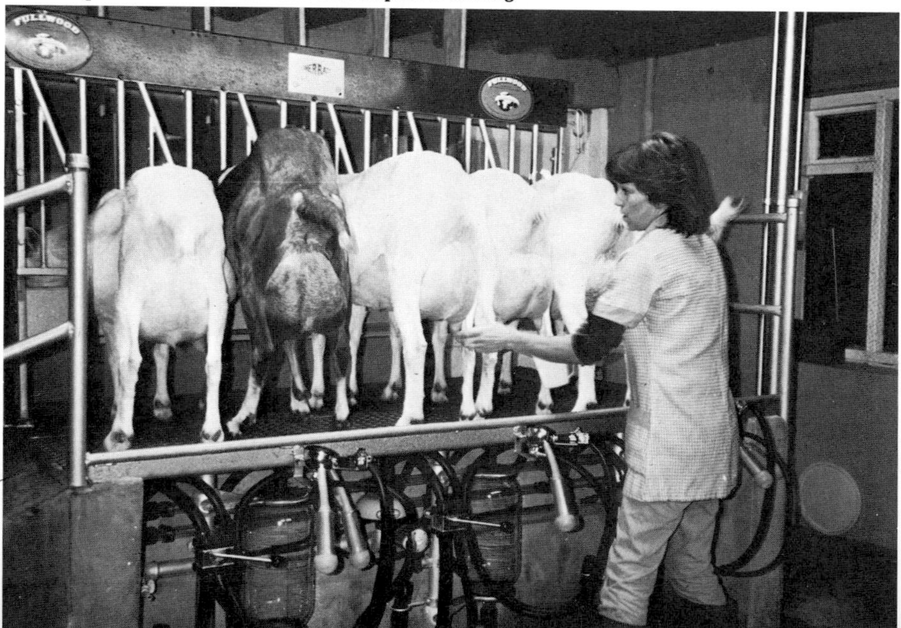

If a machine-milker is to be used on any scale, it is important to seek the advice of the manufacturer. Milking units for cows are not suitable for goats because the teat clusters are different in size. The vacuum pressure is also higher on milkers for cows. If used on goats, the teat clusters may be drawn too high, pinching the area above the top of the teat canal, and causing damage to the udder.

Hand milking This is not difficult to learn but it does take practice, particularly to accustom the muscles at the base of the thumb and wrists. They are likely to suffer considerably for the first few weeks. The best way of learning is to attend a course on goat management, or to help out on someone else's goat farm.

The basic procedure is to hold the teats in such a way that the top of the index fingers and thumbs close off the top of the teat canals, while the rest of the hands – the remaining fingers and palms – squeeze the bottom part of the canal. This has the effect of forcing milk out of the teat openings. After the milk has gone, the pressure at the top of the canals is released, allowing them to fill up again from the milk cistern, and the procedure is repeated. It is usual for this to alternate with the two teats, so that a rhythm of milking is established.

On no account should the teats be pulled down. This is the most common error in hand milking. The hands should remain at the same level throughout. Another frequent fault is to pinch off the top of the teat canal too high so that the tissues above are hurt. Any goat, no matter how even-

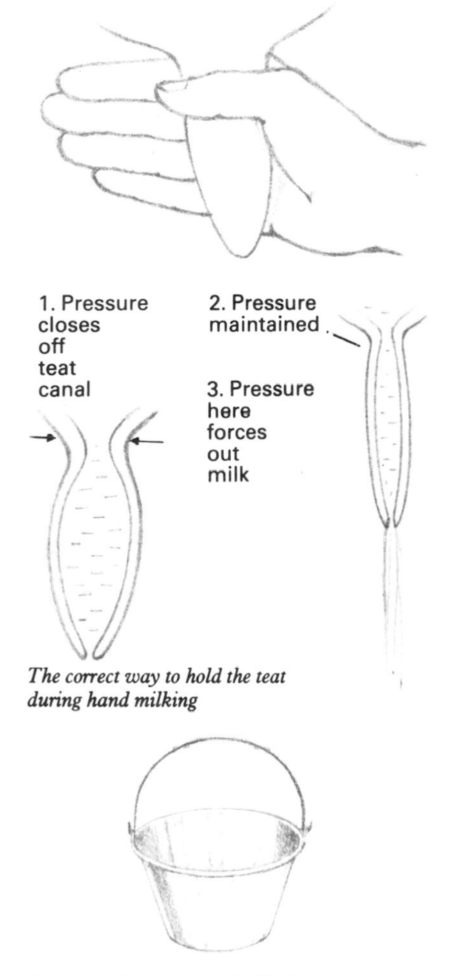

The correct way to hold the teat during hand milking

A short bucket is more suitable for goats than the normal size. Stainless steel is the most hygienic and easy to clean

tempered, is likely to react to such treatment, usually by plonking her foot firmly in the milking pail.

Infinite patience is required when milking a goat for the first time. A first kidder may have sore teats and using a purpose-made udder ointment to soothe them is essential.

Another aspect to bear in mind is that long nails can hurt a goat. They should be kept short and neatly filed.

Milking routine

Whatever the scale of milking, the following principles are essential, if an acceptable standard of hygienic milk handling is to be maintained, and if contented, unharrassed goats are to be the aim.

Prepare the area This will involve putting food in buckets or troughs to keep the goats contented while being milked. If they are hand-milked, the milking pail should be washed out with hot water and be supplied with a lid. Stainless steel milking buckets are widely available and are preferable to plastic ones which become scratched and more likely to harbour bacteria. Milking machines should be checked to ensure that filters are in place and that the system is operating smoothly.

Prepare yourself Wear an overall kept specifically for milking and keep long hair covered. Wash your hands thoroughly before starting.

Prepare the goats Marshal them outside and bring them in in a quiet and orderly fashion. They quickly get used to taking up the same positions every day, and it would be a foolhardy person who tried to make them deviate from their normal pattern.

Once the goats are in position on their milking stand and yoked in position, the udders should be washed and dried with clean udder

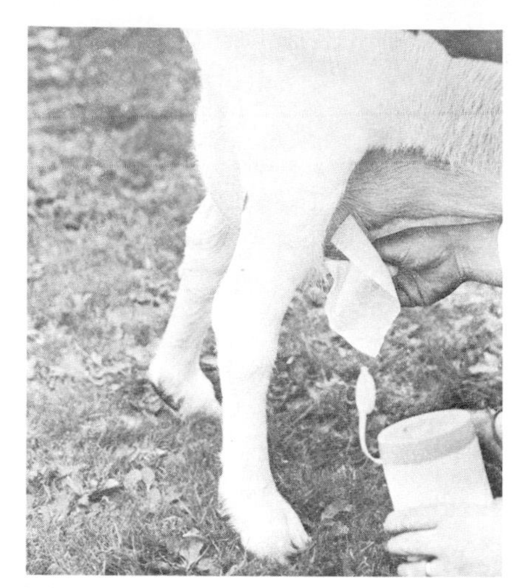

Wiping the teats before milking

cloths. Proprietary dairy cleansing and sterilizing agents are widely available for this purpose. Make sure that the udder is quite dry before milking commences.

Check the milk Check the first few squirts of milk from each teat for the presence of mastitis (see page 154). This is an infectious condition of the udder which indicates its presence in clots of milk. If you are hand-milking, use a strip cup. This is a purpose-made cup with a black, detachable interior dish. Any clots in the milk will show up against the dark surface. Pipeline units normally have mastitis detectors incorporated within them.

Milk quickly and efficiently In the previous section we outlined the correct way of milking. With machine milking, the average time is two to three minutes at a speed of around 80–90 pulsations a minute. Hand-

Pre-milking into a strip cup

milking speed will obviously vary with individuals, but is usually around five minutes to a quarter of an hour. The goat becomes used to a certain time and, if she thinks a new milker is taking longer than necessary, is likely to become fidgety.

Carry out teat dipping This is a simple procedure to ensure that the possibilities of mastitis are kept to a minimum. A teat cup containing an antiseptic preparation is held under the udder and the end of each teat is dipped in it. Once this simple procedure has been carried out, the goats are released and taken back to their yard, stalls or pasture, as the case may be.

Filter the milk The milk needs to be filtered in case any extraneous matter such as the occasional hair has fallen into it. Machine systems will have filter units in the pipeline. Hand milkers will need to do it manually. You can

Teat dipping after milking

buy small filtering units, which simply consist of a holder with disposable filter papers to fit, which is placed on top of a stainless steel bucket. You simply pour the milk from the milking pail through the filter. On a very

Filtering milk

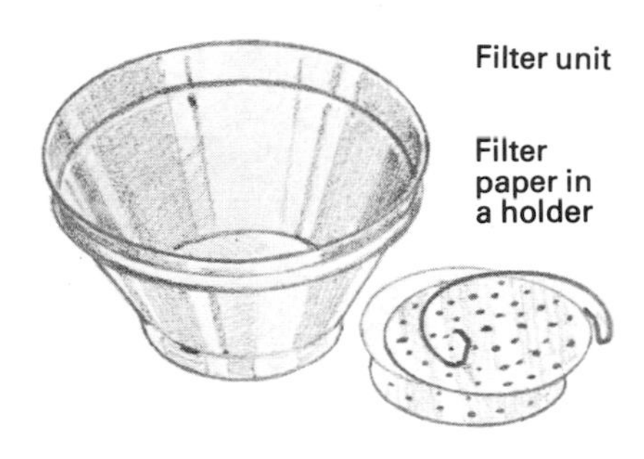

On a larger scale, a filtering unit with dairy filters should be used

small scale, where the milk is for domestic use only, a plastic kitchen strainer will suffice.

Cool the milk The milk should be cooled immediately to 5°C (41°F). A pipeline unit will have the milk going directly to a bulk tank where this is carried out automatically. On a smaller scale, an in-churn cooler is ideal. This is a hollow tube which is inserted in the milk, and which has cold water circulating through it. Where only a single pail of milk is concerned, the quickest way of cooling the milk is to place it in a sink and run cold water around it. Ensure that the lid is on so that water does not splash into the milk.

Clean the milking area and equipment The milking area should be cleaned out in case any droppings have accumulated. A purpose-built milking parlour will normally have a concrete floor which can be hosed down.

All milking equipment should be rinsed in cold water, then in hot water. Pails should then be stored upside down until needed next time. Milking machines require special procedures of washing and disinfecting, and it is advisable to follow the manufacturer's instructions.

Buckets may, over a period of time, acquire a layer of 'milk stone', particularly in hard water areas. In itself this is harmless, but it may provide an area where bacteria may become lodged. Dairying suppliers sell a purpose-made product based on phosphoric acid for removing this, as well as a dairying disinfectant.

PROCESSING THE MILK Once you have extracted the milk from the goats, it is time to take it to the milk processing area. Depending on the scale of operations, this will either be the household kitchen or a purpose-built dairy (page 40). It is here, and not in the milking area, that the filtering, cooling and, if necessary, the pasteurization should take place.

Pasteurizing

It is advisable to pasteurize all milk that is being offered for sale: not only does it keep fresher longer, but it is unlikely to develop off-flavours. Most important of all, the process destroys any possibly harmful bacteria.

On a small, family scale, pasteurization is not crucial, although possibly still worth doing, particularly if the milk supply is not likely to be used up on the same day.

Large, commercial herds will have a pasteurization unit as part of the milk handling system. On a smaller scale, small pasteurizers are available from specialist suppliers. Where only one or two goats are involved, the

Pasteurizing the home milk supply

easiest way of pasteurizing the milk is to pour it into a large saucepan, heat until the temperature reaches 82°C (180°F), and then pour into previously washed bottles or other containers. They can then be cooled by placing them in a sink of cold, running water before you remove them to the refrigerator.

Packaging and storage

On a domestic scale, any glass bottles with tops are suitable for storing milk, as long as they can be effectively cleaned with hot water. Jugs can be used in refrigerators, but containers with tops are better because there is no possiblity of developing taints from other foods in the refrigerator.

All milk should be stored in a refrigerator until used. Where this is not possible, the bottles can be placed in a container of cold water, replenishing the cold water as it warms up.

If you are going to offer your goat's milk for sale, it is essential to use purpose-made milk packaging. There is a wide range of containers for you to choose from: waxed paper cartons, plastic bags, polythene bottles and cans. Cans are widely used in the USA, though rare in Britain.

Fresh milk is best sold in cartons or polythene bottles, ideally on the day of production. Goat's milk also freezes well and for this, plastic bags are suitable. They are widely available from suppliers, along with small heat-sealing devices for closing them.

If you are going to store your supply in a refrigerator, the tempera-

A range of packaging materials for goat's milk and milk products

ture should not exceed 5°C (41°F). When it is stored in a deep freeze the temperature should stay below −18°C (−0.4°F).

Selling milk

It is not a particularly good idea to sell goat's milk unless you can be certain of producing enough milk all year round. Most goats produce more milk in summer than they do in winter. It is a great temptation to sell the summer surplus, and there is rarely a shortage of demand, but it is not fair on regular customers to let them down when winter comes. Larger-scale producers who supply milk all through the year often complain that summer sellers undercut their prices, then expect them to take over the customers when the supply runs out.

If, however, you do manage to produce milk for sale all year round, and still keep the enterprise on a small scale, you will use your time best if you concentrate on farmgate sales, plus the most useful retail outlets such as local delicatessens and health food shops.

Many customers are quite happy to come and collect their milk by car. One of the great advantages of frozen goat's milk is that a customer can restrict his visits to about once a week or once a month, taking away a bulk supply for the deep freeze for use as necessary.

Dairy sales and the law

Goat's milk must be labelled as such, indicating the amount and a 'sell-by' date, in order to meet marketing and description requirements. A wide range of labelling materials and equipment are available from specialist suppliers. Where the milk is sold through a retailer, the carton should also show the producer's name and address.

A curious and anomalous situation exists in British law in relation to goat's milk. It is not officially regarded as 'milk', because the dairying regulations state specifically that they apply only to cow's milk. (This is another legacy of the traditional disparaging attitude to goats.) Thus anyone who is a registered 'milk' producer has to comply with strict rules as to where and how he sells his produce. The 'milk' must be sold through a licensed dairy retailer, whereas the goat owner is free to sell through any retailer and indeed wherever he likes.

This situation is unlikely to continue for long, in the light of the steady rise in consumption of goat's milk and goat dairy products. Meanwhile, of course, normal health regulations which apply to all foods prevail.

Whatever the legal position, it is in the goatkeeper's interests to produce good quality, clean milk, free of taints and potentially harmful bacteria. These are things which customers and family members alike deserve.

The Code of Practice for milking is in chapter 19.

Milk recording

Recording the amount of milk each goat gives is important because it gives a clear idea which are the best

Weighing the milk enables accurate records to be kept

producers over a period of time. These are obviously the ones to take priority as future breeding stock for herd replacements or for selling.

Milk is traditionally recorded by weight because the frothiness of new milk makes it difficult to estimate the volume. The easiest system is to have a suspended spring balance scale in a convenient place. It is then simply a matter of attaching the pail on to the hook at the end of the scale and recording the weight, not forgetting to subtract the weight of the bucket.

In the interests of those who may not think in Imperial measurements, and who find it difficult to convert weight of milk to volume, the following conversions may be useful:

> United Kingdom: 1 gallon = 4.5 litres = 10 lb 5 oz = 4.68 kg
> USA: 1 gallon = 3.8 litres = 8 lb 10 oz = 3.9 kg.

A typical milk recording book would have entries such as the ones recorded below.

This system of recording is satisfactory for personal records, but is not part of an officially recognized scheme. A commercial breeder would find it more appropriate to have milk recorded by the Milk Marketing Board, or the equivalent body in other countries. It is relatively expensive, and not geared to the needs of the smaller producer. An alternative for the latter would be the Club Milk Recording operated by local goat societies who are affiliated to the British Goat Society.

Milk record book

MILK RECORD

Date	Goat	Morning	Evening	Total.
10/6/1987	Florence	5lb 2oz	4lb 13oz	9lb 15oz
	Bilberry	5lb 0oz	4lb 1oz	9lb 1oz

10. DAIRY PRODUCTS

Milk can be transformed into a variety of products, including yoghurt, cheese, butter and ice cream. Most goatkeepers operate on a domestic scale, using their kitchens as a dairy. It is only the larger, more specialized enterprise which has a separate dairy. Where milk or milk products are to be sold, then it is advisable to have such a facility. Whatever the scale of operations, absolute cleanliness and sterilization of equipment are essential.

YOGHURT Yoghurt is milk which has been curdled by the action of lactic acid-producing bacteria. These micro-organisms feed on the milk sugar **lactose**, producing lactic acid as a byproduct of their activities. The acid then acts on the milk protein **casein** causing **curdling** (the production of curds) to take place.

A small dairy specializing in goat's milk products

No one can be certain where yoghurt originated, although it is likely to have been in the Middle East or India. In hot terrains, where fresh milk would have become sour in a short time, yoghurt was a practical way of storing it for use as a drink or strained as a food. It has a long tradition of use in the Caucasus, Bulgaria, Greece and Turkey, as well as the Middle East, and many claims are made for its health-giving properties.

The micro-organisms responsible for producing yoghurt are *Lactobacillus bulgaricus*, *L. acidophilus* and *Streptococcus thermophilus*. They are available in pure form as yoghurt 'starters' from dairying suppliers. Alternatively, a little 'live' yoghurt (yoghurt from a previous batch) can be used as a starter.

Yoghurt for the family

1 pint (500 ml) goat's milk
1 tablespoon plain live goat's milk yoghurt

Heat the milk in a saucepan until it is just about to simmer. This is at 82°C (180°F) when the heat pasteurizes the milk and kills off any unwanted micro-organisms. A dairy thermometer is useful here. Now cool the milk until it reaches 43°C (110°F).

Pour most of the milk into a thermos flask which has been previously rinsed out with boiling water. Blend the rest of the milk with the live yoghurt and pour it into the flask. Put on the lid and give the flask a good shake, then place it in a warm place, such as an airing cupboard, and leave overnight.

The following day, transfer the yoghurt to a bowl, pouring off any thin surface liquid, and leave to firm in the refrigerator. Once it has cooled, it is ready for eating.

Although good results can be had from using live yoghurt as a starter, it is not always reliable. A surer method is to use a purpose-made starter. These are available from dairying suppliers and from many health food shops. They normally come in small sachets, but occasionally in liquid form, and it is important to follow the manufacturer's instructions.

Purpose-made yoghurt makers are also available for domestic use. They are not expensive and are well worth the cost. They are basically containers with a thermostatic control so that a regular incubation temperature is maintained. Failure to provide this is a frequent cause of failure in yoghurt making.

Fruit or flavoured yoghurts are easy to produce. Fresh fruit such as strawberries or raspberries should be washed, drained and, if necessary, chopped. Fold them gently into the cold yoghurt, adding honey or caster sugar to taste. If you prefer it you can use tinned fruit, pouring off most of the syrup (or juice), and leaving just enough to sweeten the yoghurt.

Yoghurt for sale

Where yoghurt is made for sale, the quantities involved are obviously greater. Although fairly large domestic yoghurt makers are available, as well as large capacity thermos flasks, a small commercial unit is more appropriate. The choice here is between a **cabinet unit** or a **vat**.

Yoghurt-making cabinet with shelves to hold trays of individual pots

The cabinet unit looks rather like a refrigerator with shelves. Milk with the commercial starter is poured into cartons which are then placed on the shelves and then incubated for several hours. The cabinet, which works rather like a refrigerator in reverse, is equipped with a thermostat to provide optimum conditions. The smallest size is normally one which produces 140 cartons at a time, and is popular with small-scale goat enterprises.

The vat-type yoghurt maker usually comes in various sizes, the smallest having a capacity of 23 litres (5 gallons). It is made up of a stainless steel bucket with lid which can be lifted out of its surrounding water jacket. The water is heated by a 3 kw heater equipped with an electronic thermostat. After initial pasteuriza-

Yoghurt-making vat to make a batch of yoghurt

tion, the milk with starter is incubated for five to six hours. After that, cooling is brought about by draining away the hot water and running cold water through the outer jacket. The yoghurt is then transferred to cartons.

Fruit yoghurts are produced by folding in the appropriate fruit pulp into the completed, cold yoghurt. Some producers may prefer to add the pulp before incubation, particularly if they are using a cabinet-type maker involving individual cartons. Pure fruit pulp for yoghurt production is available in relatively small quantities from dairying suppliers who cater for the smaller producer.

Packaging and labelling

Yoghurt produced for sale must obviously be properly packaged in attractive containers. Plastic pots are widely available and the greater the number bought, the greater the discount from the manufacturer. They are available with snap-on lids or heat-sealed foil caps. The latter will require a sealing unit, but hand-operated ones are available for a relatively low cost. The foils are usually available in a number of designs, indicating which fruit is contained in the yoghurt.

Labelling is an important aspect of marketing. In Britain, the Trades Description Act requires that the description on a carton is accurate. EEC regulations in relation to food require the list of ingredients to be clearly displayed. The USA, Australia, New Zealand and South Africa have similar laws.

The quantity of yoghurt and its description might be as follows: *150 g (5.3 oz) – Goat's milk yoghurt – strawberries – sugar – live bacteria. Sell by 07/6/88.* This information can be incorporated on a stick-on label if production is on a small scale. Alternatively, pre-printed pots are available from manufacturers, and the 'sell-by' date can be embossed on a heat-sealed foil cap.

If you are selling your yoghurt through a retailer such as the local delicatessen store, the carton should also carry your name and address; for example, *Jolly Goats, Caprine Lane, Oldtown.*

Yoghurt should be sold as fresh as possible. While awaiting sale, it should be stored in a refrigerator.

Yoghurt-making problems

We have mentioned the need for a commercial starter if reliable results are to be expected. It is essential if yoghurt is produced for sale. A common problem with some goat's milk is that the yoghurt is thin and runny. This is usually because the milk itself is thin and lacking in proteins. Ensuring that the goats have adequate levels of protein in their feed rations is important, but it is not always the answer. Some strains of goats, particularly those which produce large volumes of milk, may not necessarily yield good quality milk with the required level of proteins. The Anglo-Nubian breed, for example, is often the choice over the British Saanen (for yoghurt production) because its milk is thicker.

One way of overcoming the problem of thin yoghurt is to add a little powdered milk to the original milk before incubation. Dried goat's milk is available from specialist suppliers. If the yoghurt is for sale, the addition of powdered milk must be declared on the packaging.

The traditional way of overcoming the problem was to boil the milk for a time, driving off a proportion of the water so that, when cooled, it provided a thicker medium to start with. The expense in heating and time costs makes it hardly worth pursuing, not to mention the loss in overall nutritional value.

Finally, it is important to remember that any goat which is being treated with antibiotics for a condition such as mastitis will produce milk containing antibiotic residues. The milk must not be offered for sale, nor should it be used domestically. Such milk will not produce yoghurt anyway, because the antibiotic will kill the lactic-acid producing bacteria.

BUTTER Butter is produced by beating cream until the fat droplets coalesce (merge together). Anyone who has used a food mixer to beat double cream until it is stiff knows that if the mixing goes on too long, the cream congeals in a greasy mass. This is butter.

It is not usual to make goat's milk butter because, unlike cow's milk, there is comparatively little butterfat and it is distributed throughout the milk as small globules. This makes it difficult to skim off the surface. Nevertheless, it is possible, particularly if the goats come from a line of quality milk producers with a higher butterfat level than usual. It is not a viable proposition commercially, but is worth attempting to make your own butter for family use.

Leave the milk to stand overnight in a wide, shallow saucepan. It should be covered and left in a cool place. The following day, heat it very gently, when more cream will rise to the surface to join that which has already collected overnight. Turn off

Shallow container with skimmer to ladle cream from the surface

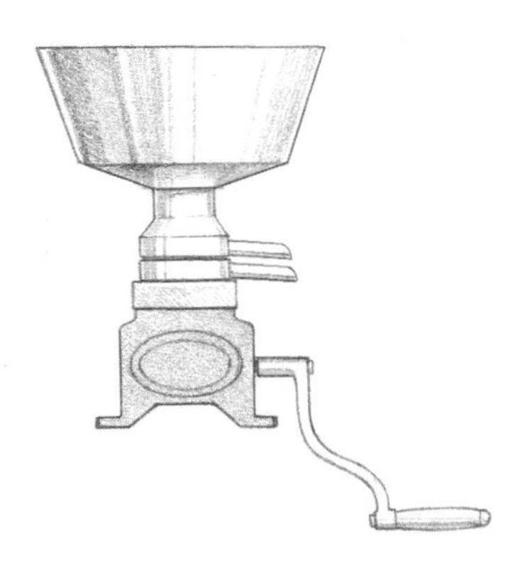

With a centrifugal cream separator it is possible to extract far more of the cream

the heat and skim the cream from the surface with a perforated ladle.

Do this with several batches of milk over a period of days until there is enough cream to churn, storing the cream in the refrigerator in the meantime.

Anyone with a cream separator is in the position of being able to separate far more of the available cream. It operates by using centrifugal force to separate the cream from the buttermilk. Small hand-operated or electrically-powered separators are available from dairying suppliers.

Small hand-operated butterchurns are also widely available. The most common type has a glass dish with a screwtop lid which incorporates an external handle for turning the internal paddle. (An alternative is to use any large glass jar with a screwtop lid and shake it vigorously, although this is quite a chore. A food mixer can be used but the problem here is that the cream shoots all over the place unless a comparatively small quantity is made at a time.)

Butter is made by half filling the churn with cream and turning the handle until particles of butter are formed. Once this 'breaking point' has been reached, pour off the liquid buttermilk and leave it to one side for later use. (It can be used for making scones.) Now add cold water to the churn to rinse away the last traces of buttermilk.

Empty the butter into a kitchen colander lined with butter muslin or close-weave netting which has been previously boiled, and squeeze out the rest of the liquid. Sprinkle on a little salt to taste. Form the butter into a pat and place in a dish. Store in the refrigerator until ready for use.

Goat's milk butter tastes exactly the same as cow's milk butter, but the appearance is quite different. It is completely white, resembling lard in all but taste. A nice way to serve it at the table is to place some fresh marigold petals on the top of the pat. These should be from the pot marigold, *Calendula officinalis*, not from the decorative French or South African marigolds.

A small hand-operated butter churn with paddles for churning the cream

CHEESE Cheese is formed when **rennet** separates the milk into **curds** and **whey**. The liquid whey is poured away while the solid curds coalesce into a mass and ripen into cheese. Soft cheeses are unpressed and have a shorter ripening period than pressed cheeses which may require several months to achieve the right degree of ripening.

Soft cheeses are quick and relatively easy to make, and require little in the way of equipment, unless they are being made for sale. Pressed cheeses need more equipment, including a proper cheese press, and are more complicated and time-consuming to make.

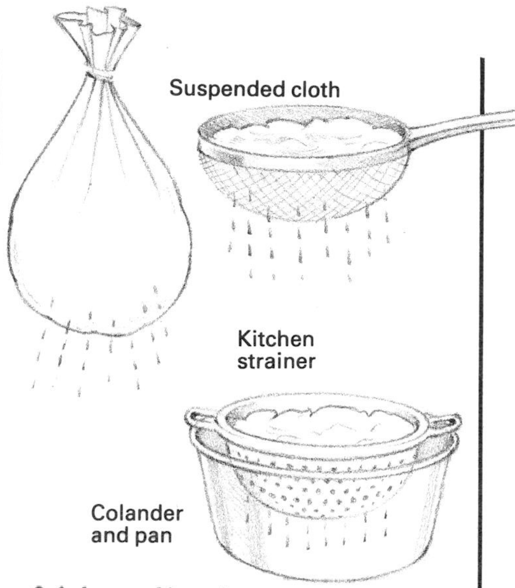

Soft cheese making at home

Lactic goat's cheese

This is an easy soft cheese to make on any scale. The quantities can be adjusted according to the amount of milk used. If made for sale, a commercial starter should be used.

10 litres (2 gallons) fresh goat's milk
2 teaspoons commercial starter or 1 small carton of plain goat's milk yoghurt
½ teaspoon of rennet
1 teaspoon of salt

If the cheese is to be sold, the milk should be pasteurized by heating to 82°C (180°F) and cooling to 24°C (75°F). For domestic use, pasteurization is not necessary, as long as the milk is clean and fresh. Add the yoghurt or culture starter when the temperature of the milk is 24°C (75°F) and stir it in.

Cover the milk and leave it in a warm place for two hours. Add the rennet, cover the milk again and leave in a warm place for 24 hours.

Next day stir up the curds which will have been formed, and drain off the liquid whey. Suspend the curds in a cheese cloth (previously boiled) and leave to drain for a couple of hours. After this open the cloth and mix the curd, sprinkling on the salt.

Leave to drain again for 24 hours, when it will be ready for eating, or for packaging and selling. Where only small quantities are involved, the draining could take place in a plastic kitchen strainer.

Soft goat's cheese

This is another straightforward recipe which can be adapted in many different ways. For example, it can be rolled in crushed peppercorns, or have chopped chives incorporated. It can also be made into a blue-streaked cheese or a white mould covered one, depending on the cultures used. Quantities can be adapted according to the amount of milk used.

10 litres (2 gallons) fresh goat's milk
½ litre (¾ pint) of previous day's milk scooped from the surface or one teaspoon commercial starter
½ teaspoon rennet

If the cheese is to be an unpasteurized one, follow this procedure: heat the milk to 20°C (68°F) and add the previous day's milk which acts as the starter culture. Cover and leave for an hour in a warm place. Add the rennet, cover the milk and leave for 24 hours in a warm place.

The next day ladle the curds, breaking them as little as possible, into plastic or stainless steel cheese moulds. These are purpose-made containers with holes in the sides and bottom, allowing drainage of the whey to take place. They are widely available from dairying suppliers.

When each mould is filled, leave them to drain for two days. At the end of this time, each cheese will have shrunk to half its size and will be firm enough to remove from the mould. Sprinkle salt over each one and place on draining mats in a cool larder for about one week. During this time, they should be turned twice a day.

If crushed peppers are to be used, they should be pressed on after the salt has been sprinkled on. If chopped chives are to be added, these should be incorporated in between the layers of curd as they are being ladled into the moulds.

If the cheeses are to be sold, it is best to pasteurize the milk and to use a commercial starter in place of the previous day's milk. The amount used will depend on the strength of the culture. Follow the manufacturer's instructions on this. These days, it is also possible to buy combined starter/rennet cultures which can save a considerable amount of time.

Blue-streaked soft goat's cheese

The ingredients and method are identical to the last cheese, except that a blue mould culture of *Penicillium roqueforti* is added at the same time as the starter. Specialist dairying suppliers sell such cultures for the small-scale cheesemaker.

White moulded soft goat's cheese

Again, the ingredients and method are identical, save for the fact that a white mould culture is sprayed on at the time the cheeses are taken out of the containers. This culture is normally made up of the two species of mould, *Penicillium camemberti* and

Cutting the curd

Penicillium candidum. It is available from dairying specialists.

Pressed goat's cheese

This takes longer to make and requires a longer ripening period before it is ready to eat. Unless production is on a fairly large scale, it is probably not economic to sell it, but it is delicious for the family table. It is paler yellow than Cheddar cheese but has a similar taste.

10 litres (2 gallons) fresh goat's milk
1 teaspoon commercial starter
½ teaspoon rennet
2 level teaspoons salt

Pasteurize the milk by heating to 82°C (180°F) then cool to 30°C (86°F). Add the starter and stir well, then leave for an hour. Add the rennet and leave for another hour, by which time the curd will have formed. Cut the curd at 1 in (2.5 cm) intervals, then at right angles to the same dimensions so that it is divided into squares. Now, using a long palette knife or the ladle, cut the strips so the result is cubes of curd floating in the whey.

Now increase the heat slowly, stirring the curds to keep them moving. The temperature should have increased to 38°C (100°F) by the end of half an hour. Turn off the heat and allow the curds to settle at the bottom before pouring off the whey.

Strain the curds in previously boiled cheese cloth and tighten the cloth as more whey drains away. Leave to drain for about an hour then open the cloth and cut the curd into four wedges. Leave to drain for a further half an hour, then break up the curd into pieces about the size of a walnut.

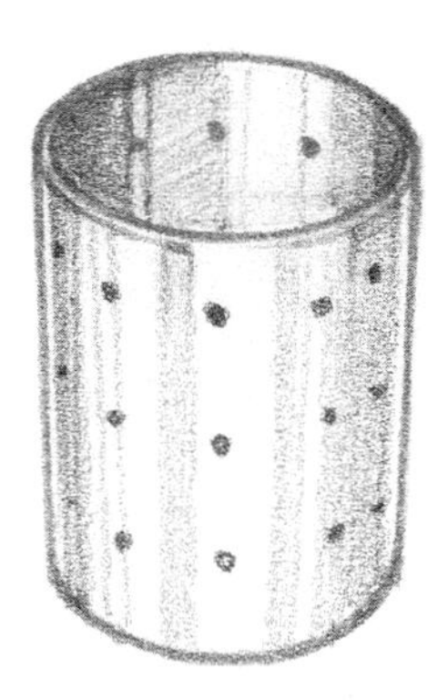

Cheese mould for soft cheese making

Sprinkle on the salt, then pack the curds into cheese moulds lined with muslin. Fold the cloth over the top, then put the wooden follower on top before placing the whole thing in the cheese press. Apply a light pressure so that whey begins to drain. An hour later increase the pressure so that more whey appears.

The following day, increase the pressure again and leave for 24 hours. At this stage remove the cheese, wash the mould and replace the cloth with a clean one. Put the cheese back, the other way up, and apply pressure again for a further 24 hours. Remove the cheese from the press and leave to dry for a few hours.

At this stage, the cheese can be rubbed with a little salt and cooking oil, bandaged with butter muslin, or covered with cheese wax which is available from dairying suppliers. It should then be stored in a cool larder and turned once a day. It will be ready to eat in about six weeks, but longer

CHEESEMAKING PROBLEMS

Problem	Remedy
Tastes too acid	Too much starter used or milk left for too long before rennet added: adjust next time
Too rubbery	Too much rennet added or temperature too high: adjust next time
Tastes bad	Sterilize equipment properly
Fermented taste	Contamination by yeasts. Keep fruit and breadmaking utensils away from cheese
Tasteless cheese	Use a commercial starter and leave to ripen properly
Too dry and crumbly	Too little rennet or curds cut too small: adjust next time
Too moist	Inadequate drainage or pressing: increase
Cheese full of holes	Storage temperature too high: reduce check also for cheese mites or mice

storage will produce a riper, more mature-tasting cheese.

Selling cheeses

Cheeses which are attractively packaged are more likely to sell than those which are not. Soft cheeses can be placed in shallow plastic cartons with lids, or in shallow wooden boxes which are sold by dairying specialists. Labelling requirements call for the weight and ingredients used to be declared as referred to in the section on yoghurt.

ICE CREAM Small quantities of ice cream can be made at home with the minimum of equipment. It is not an activity which can be carried out at a commercial level without a considerable capital expenditure. Health regulations (worldwide) are also stringent for ice cream sales, and take the matter beyond the scope of this book.

Domestic ice cream makers are available from dairying suppliers, but all that is necessary is a food mixer, a shallow dish and the freezing compartment of a refrigerator (or a deep freeze). The following recipes are easy to make and are popular with adults as well as children.

Vanilla goat's milk ice cream

240 ml (½ pint) fresh goat's milk
140 ml (6 fluid ounces) fresh goat's cream
4 egg yolks

Adequate draining, drying and ripening are essential in the cheesemaking process

75 g (3 oz) caster sugar
Few drops vanilla essence

Beat the egg yolks with the vanilla and sugar. Heat the milk until almost boiling then add gradually to the mixture, stirring continuously. Return the mixture to the heat until it thickens, still stirring, but do not allow to boil otherwise it will become stringy. Cover and leave to cool. When quite cold, fold in the cream and place in a shallow dish.

Put the dish in the freezing compartment and leave for half an hour. After this period, remove and stir thoroughly to prevent large ice crystals forming. Replace in the freezing compartment until firm. Place in room temperature for about ten minutes before serving.

Serve the ice cream plain, decorated with chopped nuts or fruit, or with chocolate sauce.

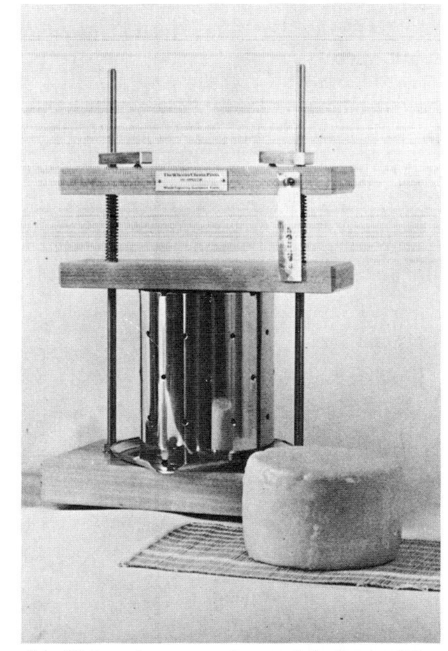

An efficient cheese press is essential when making hard cheese

Fruit goat's milk ice cream

240 ml (½ pint) fruit puree (ideally made from fresh fruit put in a food blender)
240 ml (½ pint) fresh goat's milk
140 ml (6 fluid ounces) goat's milk cream
150 g (6 oz) caster sugar
15 g (½ oz) gelatine
1 teaspoon lemon juice

Dissolve the gelatine with three tablespoons of water in a bowl over a saucepan of hot water. Add it to the fruit puree and stir well. Beat the milk and sugar, gradually adding the fruit puree and gelatine, followed by the lemon juice. Gently fold in the cream, then freeze for half an hour in the freezing compartment. After this time, remove and stir well, as with the previous recipe, and replace until frozen. When serving, decorate with fresh fruit.

CONFECTIONERY A review of products that are possible with goat's milk would not be complete without a mention of sweets. The following recipe makes attractive gifts, especially when the fudge is nicely presented. A waxed paper trifle case covered with cling foil is perfect.

Goat's milk fudge

2½ cups sugar
1 cup fine oatmeal
1 cup peanut butter
1 cup chopped nuts
½ cup goat's butter
1 cup goat's milk
6 level tablespoons cocoa
2 teaspoons vanilla

Toast the oatmeal under the grill for a few moments. Mix together the sugar, cocoa and butter in a saucepan over a gentle heat, adding the milk and vanilla. Bring to the boil, stirring continuously and cook for two minutes. Remove from the heat and stir in the peanut butter until smooth. Stir in the oatmeal and nuts and transfer the mixture to a shallow, oiled dish for setting.

When partly set, score lines halfway through the mixture with a knife to divide the fudge into squares. When quite set and cold, complete the cutting with a sharp knife.

11. GOATS FOR MEAT

The decision to rear a kid for meat is often the result of not knowing what to do with a surplus male. There is little demand for male kids, even the pedigree ones, unless they happen to be Angoras. The choice is either to kill them at birth or to rear them for a few months and put them in the family freezer. If you can harden your heart, it is worth doing.

The importation of the Angora goat into Britain has created a certain amount of commercial interest in goat meat. Although the Angora is not suited to the cool, damp British climate, its progeny from crossings with native feral goats are quite hardy. It has also been crossed with Saanens and British Saanens.

Although the main interest is in production of the quality fibres mohair, cashmere and cashgora, there is a secondary interest in the Angora's role as a meat producer. Its appearance and conformation are certainly more like those of a sheep than of a goat. It is more rectangular and more blocky than the wedge-shaped and long-legged dairy goat. The muscle tissue of the legs is heavier and its growth is more rapid in the right conditions.

So far, it has not been established with any degree of precision what these are. Research is taking place in Scotland to see if Cashmere/feral crosses have a future as exploiters of marginal land and upland farms, as producers of fibre and meat.

Commercial rearing for meat

This is a fairly well developed activity in France where I visited several farms which specialized in it. The most interesting was one which operated on a collective basis. Kids from a number of goat farms in the area were collected at regular intervals. This was done with specially adapted vans which allowed the kids to be transported in safety.

The kids were housed in communal pens, each pen having a multifeeder so that several kids could feed at the same time. The rearing was virtually the same as it would be on the small scale, except that the feeding of hay was more restricted. The kids were reared until they were around ten weeks and then slaughtered. Castration was not carried out.

The Animal and Grassland Research Institute in Britain has conducted some interesting trials in which they reared Saanen male kids and Angora × Saanen male kids. They were castrated and fed on a milk replacer diet for the first eight weeks, after spending the first four days with the mothers.

At eight weeks, an ad-lib concentrate diet was introduced, where the kids could help themselves at any time. The ration was a pelleted concentrate of barley, soya meal and fish meal with 16% crude protein. An added ration of 150 g (5½ oz) of hay per week was given.

The intake of the Angora/Saanen crosses was lower than that of the Saanens and they did not perform as well as expected, partly, it is thought, because they did not adapt well to the penning and feeding system. It is an indication that Angora crosses ought perhaps to be developed for hardiness so that they can be reared in more natural and outdoor conditions.

The Saanens performed well, putting on less fat than sheep, a factor which is important in these health conscious days.

Rearing for the home freezer

The male kid should be left with his mother for the first four days, along with any sisters he may have. Like them, he must receive the early colostrum to provide him with the antibodies to resist infection. After four days, he and his sisters and brothers, as the case may be, are taken from the dam and put in their own pen.

Kids never seem to mind this separation for they have each other's company, but the mother may bleat in distress for a day or two. If, from her stall, she can see other goats she soon settles down. Goats seem to be able to adapt to anything as long as they can see that they are not alone.

The kids' communal pen will need a thick layer of straw for bedding, and it should be well protected from draughts. If it has gaps in the walls, rather than solid board partitions, it is a good idea to place straw bales around it as insulation. Unless it is cold, they are unlikely to need a heat lamp, for they will snuggle up to each other to keep warm.

Goat kids being reared for meat in France. Here, one is being taught how to use the feeder

A drinker should be available, for it is never too early for them to learn how to drink water. But for the first week or so, their abiding interest is likely to be their bottles. Kid milk replacement rations can be used, either in individual bottles or in a lamb-bar for communal feeding.

Male kids will need more milk than the females. Initially they will have four feeds a day, reducing to three as their consumption goes up. A concentrate ration can be introduced in the second week, along with a small amount of hay.

Although it is in the interests of the goatkeeper to feed as much hay as possible to the females (who are future milk producers) too much hay at this stage is not a good thing for the meat kid. The emphasis should be on protein – milk and concentrates. However, I feel that humane considerations are important and if giving a kid the hay and roughage that he wants is going to produce less meat, then so be it!

Once he gets to the age of eight weeks, his concentrate ration can be boosted. A barley based ration is a suitable one and a possible mixture is:

2 parts barley flakes
1 part soya bean meal
1 part flaked maize

Alternatively an ordinary coarse goat ration can be given, although the milk should be continued. Hay must be available, although in relatively small quantities.

Sexual activity will begin to appear at around 12 weeks. With it will come the inevitable tainting of the meat with the smell of male goat. Unless he was castrated at a few days old, he should be killed at ten weeks. Castrated males can be killed at six months old, when they will be quite a size. It is up to you to decide whether the extra cost of feeding is worth it. It is not recommended that an uncastrated male is left to this age before slaughter. The meat is so tainted with his smell that, in my view, it is inedible. Perhaps you have a different view.

Killing is not a pleasant subject, but it must be faced. If the meat is for home consumption only, and not sold, the kid can be killed at home. If the meat is to be sold, it must be taken to an abattoir where the meat will be subject to inspection.

For the home freezer, it is obviously best and cheaper to arrange for the killing to be done at home. A local butcher may do it for you, and this is really the best option. Killing must be instantaneous, with no suffering involved.

If you are not able to do it yourself, do not attempt it, but get someone experienced to do it for you. The ideal method is to use a humane killer between the eyes. In Britain the possession of a humane killer requires registration with the police.

The easiest way to skin the kid is to suspend it, head downwards after the

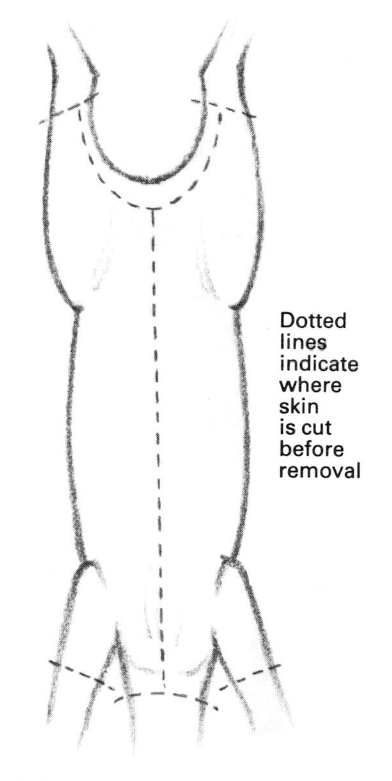

Skinning the goat

head has been severed. Alternatively, place the carcase on its back on a firm surface. Peel back the skin, breaking the connective tissue which holds it to the muscles with your hand as you go. Make a circular cut around the anus and carefully separate the skin. If you are going to cure it later, put it in a plastic bag and place it in the deep freeze for the time being.

The intestines will need to be removed from the carcase, so have a bucket ready. Cut down the middle of the abdomen without cutting into the gut. Extend the cut to the breast bone and as far down as possible into the pelvic area. Free the anus and, without allowing any of the faeces to emerge, put it in the bucket. It is still attached to the rest of the intestines of course. Lift out the intestines, using a knife if you have to, to separate the gullet. Put the whole lot in the bucket.

Now cut through the diaphragm (the membrane which separates the abdomen from the chest area) and remove the heart, lungs and windpipe.

Cut off the forelegs, and the hind feet, unless the carcase is suspended. If this is the case, do it later when the carcase is laid flat for jointing.

Butchering

Saw through the front of the pelvis and the rib cage to open up the carcase. Remove the liver and cut out the gall bladder, without spilling its contents which can give a bitter taste to the meat. Remove the kidneys, cut them in half, taking away the membranes at the same time. Leave the kidneys to soak in salt water for about half an hour before putting them in a plastic bag and freezing them. Cut the liver into slices and either put in the refrigerator if it is to be cooked later or consign it to the deep freezer in a plastic bag.

For the purposes of cutting up the carcase, it is useful to think of it in terms of a lamb, although it is longer and leaner, unless it is an Angora male kid. Saw through the back and use a chopping cleaver to separate the two sides of the carcase. If you feel like a

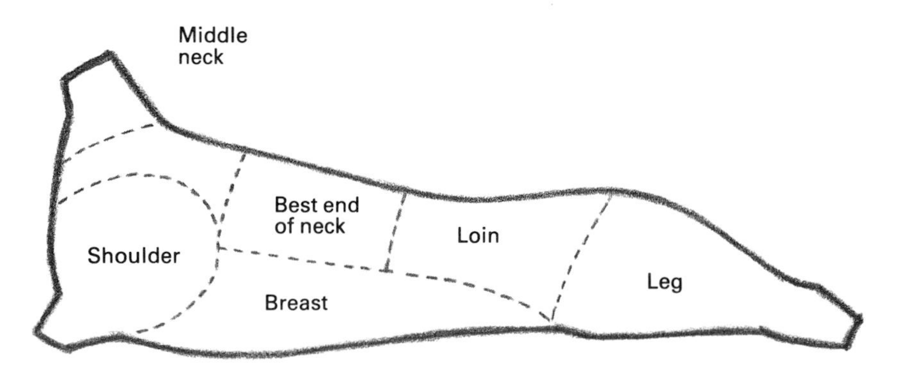

Butchering: the cuts

rest, now is the time to have it. Hang up the two sides and go and put the kettle on.

Hanging does tenderize meat, although a kid this young will be tender anyway. An older kid can be left to hang for a day or two, as long as this is in cool, fly-free conditions.

Jointing

The diagram shows the appropriate cuts to make. The joints are the ones usually associated with lamb, but I have found this to be a perfectly good way of cutting up a kid carcase. The job is made much easier if you have really sharp tools. Place each joint in a plastic bag, then seal and label it before putting it in the deep freezer. That evening, cook some of the fresh liver with onions. It will be delicious!

GOAT MEAT RECIPES Again, it is useful to think of goat meat (or chevon as it is referred to in the USA) as lamb. Most recipes which use lamb can be adapted for kid. Here are some which I have found to be popular with my family.

Greek-style goat

One leg of kid meat
1 garlic clove
1 teaspoon dried oregano
4 tablespoons olive oil
Juice of half a lemon
Salt and black pepper to taste

Place the leg in a baking tin and rub it with the garlic. Crush the clove and sprinkle it over the leg. Sprinkle on the salt, pepper and oregano. Mix together the olive oil and lemon juice and pour over the joint.

Roast in a medium oven, removing after half an hour to baste the joint with the liquid. Cook until tender and browned.

Serve with new potatoes and peas cooked with a sprig of mint in them.

Curried goat

Approx. 1 kg (2 lb) kid meat cut into cubes
1-2 tablespoons curry powder (depending on taste)
2 tablespoons flour
4 tablespoons plain yoghurt
2 cloves garlic
1 medium onion
1 small green pepper
2 tablespoons butter
1 tablespoon olive oil
Salt and pepper to taste
2 cups of water
1 teaspoon lemon juice

Mix the flour, curry powder, salt and pepper together and roll the cubed meat in the mixture. Chop the onion and green pepper and cook them in the oil and butter. When they are cooked, add the cubed and floured meat, turning the cubes over until they are browned all over. Make sure the flour does not stick to the bottom of the pan and burn.

Now add the water and lemon juice, stirring carefully to ensure that everything is well mixed. Crush the garlic and sprinkle into the mixture. Leave to cook on a slow heat until the meat is tender. Stir in the yoghurt for the last ten minutes of cooking. Serve with rice and fresh watercress.

This recipe is particularly suitable for older goat meat which may need long slow cooking. If the meat is from an animal older than a year, the cubes of meat should be stewed in water first, before the recipe is followed, otherwise it will be like the proverbial old boots.

Goat sausages

Approx. 1 kg (2 lb) kid or goat meat
110 g (4 oz) suet
2 level teaspoons salt
110 g (4 oz) breadcrumbs
12 fresh sage leaves or 1 teaspoon dried herbs
Black pepper to taste
Sausage casings

Mince the meat and lard thoroughly. Chop the fresh sage leaves and sprinkle into the mixture with the salt and pepper. Stir in the breadcrumbs and mix well. Stuff the casings with the mixture (casings are available from specialist suppliers). Alternatively, add a beaten egg to the mixture and shape it into sausage shapes with well-floured hands.

CURING A GOAT SKIN That kid skin which was hurriedly confined to the deep freezer while the butchering and jointing was going on can now be cured. If it is in the freezer, there is no particular rush. It can stay there for a couple of months, but it will be taking up room.

There are several ways of curing a skin, some more difficult than others. The following is probably the best because it uses the same formula that many professionals use.

Allow the skin to defrost completely, then wash it well in warm water and soap flakes. If it is particularly soiled, leave it to soak in cold water for about an hour first. After washing, rinse it well and put it in the spin dryer. While it is spinning, prepare the tanning compound as follows:

1 part aluminium sulphate (alum)
1 part cresolurea formaldehyde
1 part salt

The first two chemicals can be obtained separately from a chemist, or as a combined mixture from specialist suppliers, under a variety of trade names. The salt can be the ordinary household variety. The amount of mixture made up will depend on the number of skins to be tanned.

One goat skin will require 700 ml (1½ pints) of liquid compound initially. This is achieved by adding 225 g (½ lb) of each of the three ingredients to 700 ml (1½ pints) of water. Make up proportionally more for each extra skin.

Exercise care when making up the solution, for it will froth up a bit, and splashes on the skin should be avoided. Wear rubber gloves and mix the ingredients in a plastic bucket, not a metal one.

Lay the skin flesh side up on a work surface and scrape away as much as possible of the tissues. Find a knife, blade or scraper which is comfortable to use. At this stage it will be difficult because the skin is so slippery. Rub in half of the mixture, then leave overnight. The next day rub in the rest of the mixture.

It is much easier if several skins are done at once because you can make up enough mixture to put in a plastic dustbin with a lid, and just leave the skins to soak for several days. As the tanning process proceeds, the tissues which would not come away with the initial scraping will have gone hard so that they are more easily scraped away.

If you are doing one skin only, you will need to make up a second lot of the mixture. Rub this in over the next three days, scraping at the skin each time. At the end of this period, the skin will have turned white, indicating that the tanning is complete.

Now mix 28 g (1 oz) of sodium bicarbonate in 140 ml (6 fluid ounces) of water and spread it over the flesh side of the skin. (Bicarbonate of soda is available from most supermarkets.) Leave the skin overnight. Next day, wash it well with warm water and soap flakes. Put it in the spin dryer and then hang it on the line until nearly dry.

When it is nearly dry, but not quite, rub in some neatsfoot oil (available from saddlery shops) and stretch the skin in different directions. Now, give it a good rub all over with sandpaper and a sanding block. This will raise the surface and give it a suede look. Give the hair side a good brush, trimming away any straggling bits around the edges. The skin is now cured!

HORN BUTTONS Meat goats will not have been disbudded as kids, so there will be horns from the carcases. If you have the necessary equipment, it is possible to make buttons out of them. You need an electric drill with a saw attachment and a fine drill.

Secure the horn in a clamp and saw it up into transverse sections. The thickness can be whatever you decide, bearing in mind that the products are to be buttons, not millstones. Now take each section and drill two holes in the centre. Give each one a rub with sandpaper to smooth the edges. Home knitters love them, particularly on chunky knitwear.

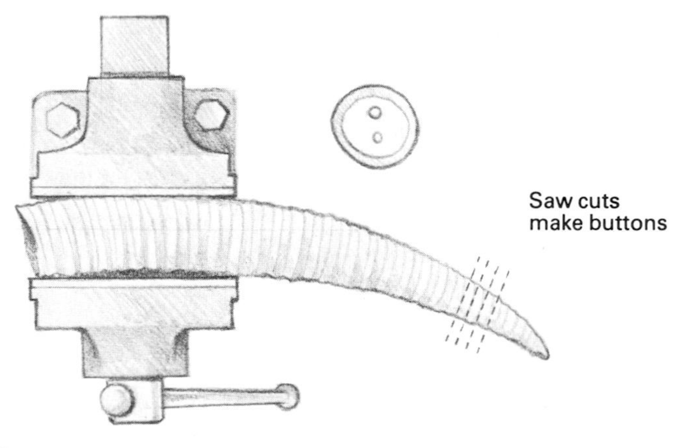

Making horn buttons

12. FIBRE PRODUCING GOATS

The fibres which are produced by goats are **cashmere**, **mohair** and **cashgora**. The first two are well known, and have been for thousands of years. Mohair from the Angora goat was valued in Biblical times, while cashmere from the hardy breeds of the cool, mountainous areas was treasured by the Caesars of ancient Rome.

Cashgora is a brand new fibre produced by crossing Angora males with any other breed of goat, but particularly with those that produce a certain level of cashmere. It may be regarded as a half-way stage between the first two fibres, although this is an over-simplification.

As there can be confusion about the precise differences and origins of the fibres (especially as there is also angora wool which comes from the Angora rabbit), it is worth looking at them in turn.

CASHMERE There is no specific breed of goat called a Cashmere. The term is applied to a goat of any breed or type which produces substantial amounts of cashmere fibre.

A goat's coat is made up of **primary hairs** (which are also known as **guard**

Cashmere-producing goats

hairs) and a finer, shorter **undercoat** which is composed of **secondary hairs**. In a normal dairy goat, the primary hairs are long and coarse, and form the greater part of the coat. Some goats, in cold areas, produce more of the fine underhair in order to keep warm.

Primary fibres are hollow (medullated); both primary and secondary hairs are produced from groups of **follicles** in the skin. If the secondary hairs are less than 19 micrometres (thousandths of a millimetre, until recently known as microns) in diameter, they are known as **cashmere**.

A Cashmere goat is one which has a particularly thick secondary coat. Any goat will produce some cashmere in winter, even an ordinary dairy goat. Selective breeding has produced goats with annual yields of 200-300 g (6-10 oz), while the best animals produce up to 600 g (1¼ lb).

Cashmere is the finest and lightest fibre produced by goats. The spun and woven fabric is warm, light and remarkably tough. It is the most exclusive and expensive of the natural fibres and is used in knitwear, underwear, coats, sportswear, suits, rugs, scarves and even carpets where it imparts softness to the product.

The major exporters of cashmere are China, Mongolia, Iran and Afghanistan, all areas which have rugged, mountainous and exposed areas, and where goats produce a substantial undercoat to keep warm. China is now processing much of its own fibre and exporting cashmere yarn and textiles; however, instability in the other cashmere producing countries has resulted in a strong world demand over supply.

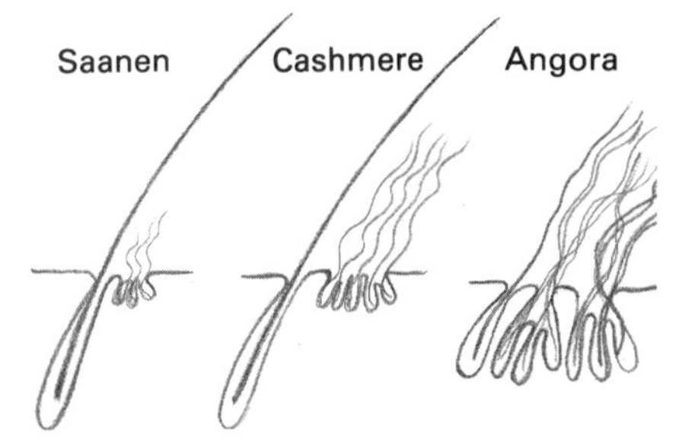

Three types of coat composition

The world's major processing company for cashmere is Dawson's International of Selkirk in Scotland. It buys in 50 kg (120 lb) bales or bundles. Small-scale goatkeepers cannot produce anything like this amount at a time, but are able to sell it through a cooperative which pools the small amounts into the minimum lots required. Such a scheme is operated by the Scottish Cashmere Producers' Association for its members.

The prospects for cashmere production in future look promising. A project at The Hill Farming Research Organization in Edinburgh has been studying the effectiveness of goats for improving the most inaccessible sheep pastures in Scotland. The goats have been found to have a beneficial effect on these upland areas, eating scrub, sedges and rushes, which sheep cannot eat, and thereby making pasture more accessible to sheep.

The HFRO research team also set out to see whether the goats would be best suited to fibre or meat production. The result of their extensive experimental work is that they have succeeded in creating a reasonable Scottish cashmere-producing goat.

CASHMERE PRODUCTION OF MALE KID GOATS

Breed of goat	Live weight (kg)	Estimated weight of down (g)	Thickness of down fibres in micrometres (= thousands of 1 mm)
Anglo-Nubian	18.3	5.9	13.2
British Saanen	24.3	26.9	15.4
British Toggenburg	25.6	63.6	13.6
Feral	13.6	37.3	14.7
Feral × Anglo-Nubian	17.6	8.6	13.6
Feral × B. Saanen	21.1	42.3	15.7
Feral × Toggenburg	19.5	29.5	14.5
Toggenburg × Feral	17.5	64.2	15.9
Angora × Feral	14.4	147.5	18.7

Source: Hill Farming Research Organization, September 1984.

Breeding for cashmere

Breeding for cashmere should aim for the following standards:

Fibre diameter The average fibre diameter should be 15 micrometres. The finer the fibre, the higher the value.

Colour White has the highest value. Grey, followed by brown, earn less.

When breeding for cashmere fibre, aim first for weight of cashmere produced. Colour and fineness can be developed second. Larger goats have a larger surface area for fibre production and males produce more than females. The weight of fibre and fibre diameter increase with age, and longer fibres provide the greatest weight.

If you have suitable land and are considering goats for cashmere production, you can purchase female, in-kid goats carrying cashmere-producing kids from an embryo transfer programme. The number of animals for sale in Britain is still fairly limited, so an alternative may be to buy goats which produce some cashmere, and then breed selectively from them.

Feral goats tend to have more cashmere than normal domestic goats. If you can purchase a good cashmere-producing buck, or make use of a stud, this will speed up your breeding programme considerably.

MOHAIR Mohair is the name given to the fibre from the undercoat of the Angora goat. Angoras have a different kind of coat from other goats. It consists of very few primary hairs, which are known as the **kemp** and are coarse, and the mass of the coat which is made up of long secondary fibres that are between 23-38 micrometres in diameter, in other words, the **mohair**.

Mohair is coarser than cashmere, but it has valuable properties, particularly its lustre and softness. The first coat of an Angora kid consists of 45% kemp, but most of this is shed at around three months, when the secondary fibres grow. At six months, the kid has its first full fleece consisting of the finer but longer fibres which originate from the secondary follicles. This is the finest and best mohair, commanding a premium price. The coat becomes more coarse as the goat gets older.

A pure-bred Angora goat

The chief characteristic of mohair fibre is its **lustre**, which distinguishes it from wool and other natural fibres. It stems from the large cuticle scales on the fibres, and the way in which these reflect light. Mohair is often called the 'diamond' fibre for this reason.

As well as possessing this sheen, mohair is soft and resilient. It is resistant to creasing and since the fibres are long they are easily processed and blended with other fibres.

The required colour is white. Commercially, coloured fibres are not accepted, but there is a growing demand for them from craft spinners and weavers.

BREEDING AND GENERAL HUSBANDRY There are several routes into keeping Angora goats. They depend on the aim, and on the size of wallet available. At the time of writing New Zealand, the USA and Australia are already well into the field, but Britain is only at the beginning of an Angora industry.

(1) You can start by buying Angora goats, but prices (in the UK) are currently higher than can be justified by the returns gained from mohair production. As supply and demand come more into balance, and the number of Angora goats grows, prices will come down to levels which are more linked to the returns.

(2) As another means of entering mohair production, the small-scale goatkeeper can buy first, second or third generation Angora crosses at correspondingly lower prices, and work to upgrade a small herd or flock. (The term 'herd' is widely used for dairy goats, but the more commonly used collective noun for fibre goats is 'flock'. Both are correct, but as the industry increases it will be interesting to see how the usage settles down in the long term.)

If a small producer buys Angora crosses, the fibre will be of a lower standard than that demanded commercially, but he or she will be in a position to cater for the craftspeople. The coarse kemp, frowned on by the mohair world, is popular with spinners. Mohair on its own is difficult to spin, as I have discovered to my cost, because it is fine and slippery. The coarse kemp gives it just the right degree of friction when handling it and feeding it into the wheel.

A group of dairy goats with their pure-bred Angora kids obtained by embryo transfer.

(**3**) If you can afford it, buying crosses and grading up with a good, pure-bred stud Angora is an effective way of entering the field. **'Grading up'** is a term meaning a cross-breeding process which involves back crossing to a particular breed or type until the desired standard is achieved.

(**4**) If you have the necessary minimum number of goats for it to be a worthwhile proposition, a fourth way is to make use of the artificial insemination services. In this way, the top males can contribute to the improving process.

In Britain, the British Angora Goat Society registers all pure-bred stock imported into the country, raised from imported embryos or home-bred. They also administer a grading-up scheme and operate a 'fibre pool' for marketing members' mohair. The USA, New Zealand, Australia and South Africa have similar bodies,

which are listed in the reference section (page 162).

The general objective in breeding Angora goats is to produce the best quality fleece. This should be particularly apparent on the head, with little or no kemp around the horns or along the back. The fleece itself should be white, dense, lustrous and soft, with uniform growth and length.

Cashgora is a fibre produced by the progeny of an Angora or Angora cross male mated with a female of another breed which has a reasonable yield of cashmere, such as a feral goat, or improved feral. The diameter of the fibre is between 19-25 micrometres, with an ideal of below 22 micrometres.

Cashgora is finer than mohair, and shows some characteristics of both cashmere and mohair.

The effect of using an Angora male on a cashmere type female is to produce a heavier yield of secondary down fibres, which are white in colour. This has obvious advantages if the aim is to breed selectively and cross back to an improved cashmere producer. The disadvantage is that the fibre is more difficult to process.

Management of fibre goats

Husbandry for fibre goats bears some similarity to that for sheep. This is particularly true of cashmere producing goats, whose general management is similar to that of hill sheep. The Angora goat has higher nutritional requirements, more in line with dairy animals. As with sheep, the kids are encouraged to run with the mothers, instead of being separated at a few days old. In fact, Angora mothers and kids are often left in a family 'bonding' pen for a few days after birth. This is to enable the family bond to be strengthened before they are allowed to go out to pasture with other family groups.

Some people, particularly those who have a small number of Angoras or Angora crosses, prefer to bottle-feed the kids. The kids grow more quickly, but it is time-consuming and, on a commercial level, is unlikely to be cost-effective.

An adult Angora, sheared twice yearly, may produce 2.5 cm (1 in) of mohair a month. For output like this you need a good standard of nutrition. During the spring and early summer, when pasture is at its most nutritious, concentrates may not be necessary, but this will depend on the status of the individual goat.

Concentrates for Angoras should be about 16-18% protein. An in-kid and housed female, within two months of kidding, would need ad-lib hay and 300 g (10 oz) of concentrates. Around kidding, the concentrates would be increased to 400 g (14 oz). In early lactation and when spring pasture is available, she would receive 250 g (8 oz) of concentrate, with grass and hay. In late lactation, concentrates would be increased to 400 g (14 oz), with ad-lib hay and grazing. When she was no longer lactating and in early pregnancy, concentrates would be 200 g (7 oz), with ad-lib hay.

Cashmeres require concentrates in pregnancy, and in periods when the grazing is poor, otherwise they will get most of their needs from hill browsing.

Angoras need to be housed in winter to protect them from the cold and damp weather. Cashmere-producing goats based on ferals are much hardier and less likely to need housing, although shelters should be available for them.

Winter quarters should allow 1.5–2 square metres (square yards) per animal, and the most suitable is that which has adequate ventilation without draughts: the Yorkshire boarding type of shelter (page 29) is ideal. In summer, Angoras in Britain need access to rain shelters such as polythene tunnels with reinforced walls.

Kidding is the same with fibre goats as it is with dairy goats. The same precautions should be taken, with kidding taking place indoors or in individual pens. In warmer climates, kidding frequently takes place outside, although the protection of corrals is provided.

As stated in the Kidding section (page 84), pregnant women should never help at kidding times, as the goats can pass on a viral infection which can cause abortion.

Shearing

Angora goats are shorn twice a year: in late winter/early spring, before kidding (no later than six weeks before kidding), and again in the autumn, before mating. Cashmere goats are shorn once, in late winter to early spring, before kidding.

The best cashmere of all would be obtained by combing it out of the goat's coat, but this is not an economic proposition.

If a goat is not shorn, it simply sheds the cashmere in handfuls soon after kidding, but it is hardly feasible to gather it up from the hills and dales of the feral wanderings.

Shearing is the same as for sheep. The goats are rounded up and yarded, and a clean, protected area is set aside for the shearing to take place. White fleeces should be done first, with coloured ones later, to avoid getting coloured fibres in the white.

Angoras should be sheared in groups – kids, goatlings, does, bucks, keeping the fibres separate. This will simplify classifying them.

Special goat shearing combs are available for the shearers. Mohair lacks the natural lanolin of sheep's wool and as a result the shears must be used at a slower rate to avoid overheating. Shearing should obviously not take place in wet weather or when the goats are wet, and a safety circuit breaker should be used, in case of accidents and possible electrocution.

Spinning the fibres

Some goatkeepers keep a small number of fibre goats purely because they provide some very interesting spinning and weaving. The first time I tried to spin pure mohair was with an Australian fleece. It was lustrous and beautiful, but immensely difficult to spin because of the lack of lanolin.

I tried rubbing oil on my hands from time to time, but this was not satisfactory. I eventually settled for mixing it with a little sheep's wool. Such an action is no doubt heresy to mohair producers, but every spinner will sympathize with the problem.

It is possible to spin mohair, of

course, but I am not one of the skilled elite of expert spinners, with the dexterity and patience required for fine, luxury fibres. I am a basic spinner, used to chunky knitwear, where mistakes do not show up as readily.

What I did glean from my experience of spinning mohair may interest other spinners. I had to decrease the tension on the wheel because it was set for sheep's wool, which is coarser, before the fibres ran smoothly. Because they are so fine, it is easy to overspin, by twisting too much. It requires far less twist to it than the average sheep's wool. Mohair with kemp in it is much easier to spin.

I have no experience of spinning cashmere but a spinning friend tells me that it needs no carding. (Carding is the process of 'combing' fibres to make them lie in the same direction, and make them easier to spin). Tension on the wheel should be loose, and there is no need to feed in the fibres as one would with wool. All that is necessary is to hold it loosely in one hand, letting the wheel draw the fibres.

Spinners are interested in all fibres. Most would be unlikely to object to mohair with kemp in it, for it is much easier to spin if it is not too pure. Goatkeepers with Angora crosses should find no difficulty in selling their kempy mohair to craft shops, local spinning groups or specialist spinning and weaving suppliers. Those who are also spinners and knitters themselves may prefer to sell finished garments.

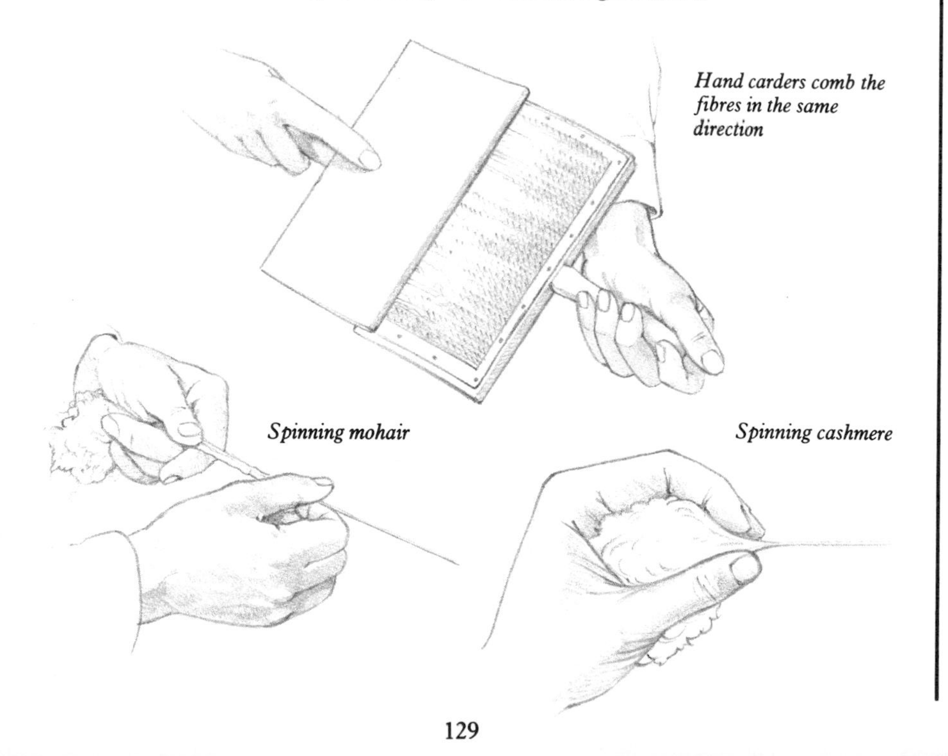

Hand carders comb the fibres in the same direction

Spinning mohair

Spinning cashmere

13. SHOWING

Many people obtain a great deal of pleasure from entering their goats at various shows and competitions. This in itself is a good enough reason for having shows, but there are other sound reasons. At a basic level, it brings together goatkeepers of varying experience who can learn from each other. Seeing other people's goats, their plus points and shortcomings, is often a useful guideline in what to aim for in the future.

On a wider perspective, shows are about setting and maintaining standards. Over the years, bodies such as the British Goat Society have sought to establish and raise the standards for goats and goatkeeping, and breeders have worked to produce better stock.

Parading at the show

There are different types and levels of goat show. The BGS, and equivalent organizations in other countries, list all shows which conform to their regulations, and which they recognize. There are also young stock shows run by local goat clubs, often with a section for children. Many rural schools have farm clubs where goats are popular inhabitants, and who are entered in such shows.

Most shows are for female goats, but there are those specifically for males, where male kids and goatlings are shown as well as the adults. These are primarily for breeders and stud owners to judge the breed points and conformation of current stud males.

Preparations for the show

The BGS, like other national goat organizations, publishes a show schedule for the coming year. With a copy of this, select the shows you wish to attend and send off for the required entry details in good time. There are different classes for different goats and it is sometimes quite a lengthy task sorting out the right classes for your particular goats.

Shows vary in size. In large ones the breeds are always separated. In a small club show, there may simply be classes for milkers, goatlings and kids. Whatever the size of the show, there are quite a few preparations.

A few words of encouragement for this British Toggenburg goat before the lineup at the show

A goat that is shown needs to be able to walk well with her owner, to stand still when required, and to be used to being tied to a post without fretting. These things require relatively long periods of training and experience, but a few trial runs before the show will jog her memory, particularly if there is a juicy titbit at the end of each session. Gentle and firm encouragement may produce results, but downright blackmail works wonders.

A few days before the show, trim her hooves. If they are not too long, a few passes of a Surform plane will suffice. It is important for a show goat to be able to walk well, so the finish on her hooves needs to be smooth and even.

The day before the event, trim any untidy hairs from her beard and legs, then give her a shampoo. Many goatkeepers use a dog shampoo, and I have used a baby shampoo to good effect, the principle being that if it does not sting a baby's eyes, it is unlikely to be objectionable to a goat. These days, purpose-made goat shampoos are available.

Dry her as quickly as possible, ideally by walking her in the sun. To prevent chilling, she should have a coat such as the home-made one illustrated, or have an old blanket put over her. There are also purpose-made goat coats on sale and one of these is smarter for going to a show, but the home-made one works well for less formal occasions.

There are preparations to be made for your wardrobe as well. A clean white overall for the line-up, wellingtons in case it is muddy on the show

A coat for a show goat

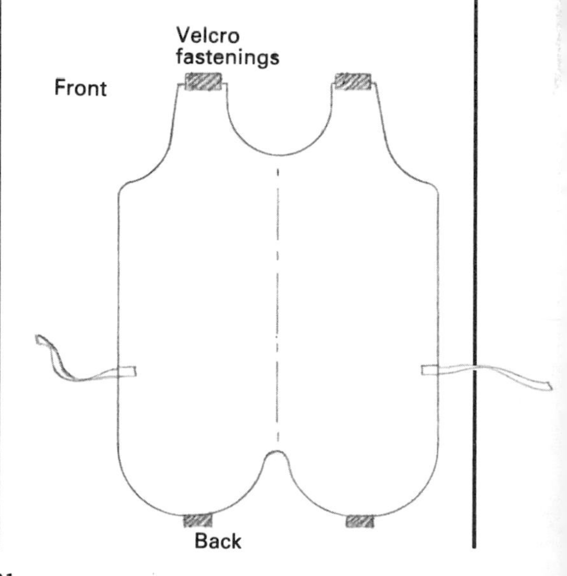

ground, and a sleeping bag for overnight stays.

Other essentials are a torch, matches, a kettle and a small stove. This is not just to produce cups of tea, although these are essential, but to heat up milk if you have young kids to be bottle-fed.

Pens for the goats are normally provided but you need to take a hay rack, pails, concentrates, some branches, collar and lead, in addition to your own essentials like a toothbrush.

Milking trials

If your goat is taking part in milking trials, it will be necessary for you to go to the show the previous evening. This is so that the stewards can strip out milk from the goats. In this way, they all start from an equal position.

The following morning, the goats with their full udders are inspected, then milking takes place. All the pails are taken to be weighed, after which butterfat samples are taken.

Details of the awards for milking trials are given in the reference section (page 165). Such trials are particularly useful in that they help to identify the good milking strains.

Health aspects of the show

Any goat which is unwell should not attend a show. By the same token, it is sensible to protect your goats from possible infection by others. Goats can be positive carriers of the Caprine Arthritis Encephalitis virus without the disease having developed. It is thought to be transmitted primarily through the milk. If you have kids, make sure that you do not give them milk from any source other than that known to be safe. Further details of CAE are given on pages 148-9.

Goat driving

Goats have traditionally been used to pull carts in many parts of the world. A castrated male is strong and sturdy, without being aggressive, and can be trained to harness. Anglo-Nubians were a popular choice in Britain before the tradition died out.

Harness goats in the past have performed various jobs: acting as water carriers, delivering local milk supplies, carting hay and straw, carrying logs, even delivering the mail.

Goat harnesses are available from suppliers, or they are not too difficult to make yourself. There is even a Goat Driving Society which puts on a spectacular display at goat shows in the summer. The participants pull decorated carts, often with children as passengers, while the accompanying walkers are dressed in traditional costume.

There is little, it seems, that goats cannot do, although one of the most unusual activities I have come across, is the lady who takes her goat carol-singing before Christmas. The goat, with some decorative glitter around her collar, accompanies the carol singers around the village and they raise money for charity. No one can resist her, but she is apparently not the most tuneful of God's creatures.

14. THE GOAT ENTERPRISE

Goatkeeping is possible on any scale, from pet animals kept for family use to the large scale commercial enterprise. Most goatkeepers are somewhere in between. Keeping any livestock is expensive. They have to be fed and looked after, and there will be veterinary bills to meet. Even if the goats are primarily pets, it makes sense to try and cover their feeding and management costs. Having a small, possibly part-time enterprise is a way of doing this.

There are many possibilities. The following is a brief look at some of them.

Selling milk If this is to be undertaken, it must be produced in a hygienic and efficient manner. Guidelines are given in Appendix 1, but it is important to check on any existing and new regulations which come into force.

The best way to sell it, from your point of view, is frozen and in bulk, with regular customers coming to buy at agreed times. Being able to supply milk in winter is essential, otherwise you will be letting customers down.

Larger enterprises will be operating on quite a different scale, normally wholesaling their milk.

Making yoghurt Specialized equipment is needed to make this on a reasonable scale, and professional packaging must be used. Local delicatessens and health food stores are the most likely retail buyers, although regular callers can also be catered for. Health regulations must be checked and met.

Cheesemaking Soft cheese is easier than a pressed cheese. It requires less specialized equipment and takes a shorter period to make. It is advisable to go on a cheesemaking course at one of the agricultural colleges if the enterprise is to be on a reasonable scale.

Some goat's soft cheese that I have tasted in Britain, the USA and Australia have been dreadful. The ideal training would be a couple of weeks on a French goat farm. One can learn more there than in a year at an agricultural college.

If this is not possible (and it is certainly unlikely) then the next best step is to read *The Fabrication of Farmstead Goat Cheese* by Jean-Claude Le Jaouen, translated from the French and published by *Cheesemaker's Journal* in the USA (see reference section).

Goat stud and boarding If you have top class, registered males then you can offer a goat stud service.

Goat boarding facilities for other goats is also a possibility. Goatkeepers need holidays, and finding someone to look after their stock can be a problem: goat boarding solves it for them.

Considerable space and good housing is needed, and you should be well insured. Do not accept any goat for mating or boarding which has not been blood-tested for CAE.

Selling goats Good quality, registered kids will always sell, particularly if they have already been disbudded, and are known to be from a good herd. The secretary of the local goat club should be kept informed of the stock you have for sale, and they can be advertised in specialist publications such as the *British Goat Society Journal* and *Home Farm* in the UK, and in *Dairy Goat Journal* in the USA (see the reference section for a list of publications).

Those with Angora and Angora crosses should find little difficulty in selling × kids, particularly to goatkeepers with an interest in fibre crafts.

Goat meat If you live in an area where there is an immigrant population from India, Pakistan or the Middle East, local butchers are likely to be interested in buying goat meat from you. Any meat sold must be slaughtered in an abattoir where it will be inspected. Details of raising kids for meat are given on pages 115-7.

Small farm park If you are within reasonable reach of an urban area, and have the space available, opening the site to the public, particularly to school parties, can be a worthwhile enterprise. The goats and kids should be in proper enclosures, with easy-to-read and attractive notices.

An exhibition showing details of the history and life of the goat makes an interesting and educational display. Demonstrations of milking, cheesemaking and yoghurt making are always popular, particularly if some of the products are for sale.

It helps to have a small picnic or cafeteria area, and a shop where

Goat kids are popular attractions at a farm park

There is increasing public demand for good quality goat products

school parties can buy souvenirs. Postcards with goat pictures go down well. There must be toilet and car parking facilities provided.

Where members of the public are coming on to your site, there is always the risk of accident. Discuss the project with your local authority and with a good insurance broker.

Goat fibre crafts Anyone who is a skilled spinner, weaver or knitter is in a good position to start a small enterprise using goat fibres. It need not be the expensive pure cashmere and mohair. The kempy mohair and cashgora from Angora crosses is cheaper and easier to use. It is possible to buy in the fleeces and process them yourself, as well as spin other people's fleeces for them.

Goat skins have an appeal as floor and wall coverings. It is possible to take orders from people who want their own goat skins cured, as well as producing your own. Reference was made to horn buttons in the section on meat goats. These are not difficult to make as long as you have a power drill with attachments. They are always popular with knitters.

Selling goat supplies Where there are goatkeepers in an area, there is always a need for supplies such as feeds, hay, hoof paring knives and medications. A small enterprise supplying some of these needs can dovetail well with your own goat-keeping.

Adequate storage space is needed, particularly for the more bulky items. Manufacturers of various products are often looking for local agents to sell their products. The important thing, particularly at the beginning, is not to over-order, otherwise you may end up with a cash flow problem.

Other goat products Once you have an existing sales area, there is a whole range of goat products available, some which you can make yourself, and others which are bought in. They include T-shirts, and jerseys with goat logos, goat decorated pottery, badges, postcards, stationery and books. The list is endless, but worth considering.

15. THE PET GOAT

Most goats are kept as family milk providers. They are usually regarded as pets, particularly if the numbers are small. But what about people who wish to keep goats purely as pets, and who are not necessarily interested in milk production or some other utility characteristic? Does the goat make a suitable pet? The answer is yes, as long as certain basic considerations are met. Perhaps the most important factor when considering keeping goats as pets is to choose those that suit the available conditions. If space is limited, it is foolish to have a large breed; a dwarf breed would be more appropriate. Where there are young children, it would be irresponsible to have a strong wilful male goat of one of the large breeds, even if it is a castrated one. Similarly, keeping a goat with horns is also inadvisable where children are involved.

Who could resist this five-week-old Saanen kid?

Any breed of goat can be reared as a pet, if suitable conditions are available, although the smaller breeds are usually the most popular in this respect. The **African Pygmy** goat is the choice of many. Its small size makes it easy to handle and it can be kept in a relatively confined space.

The **Golden Guernsey** and **Toggenburg** breeds are also popular mainly, I suspect, because of their long wavy hair. **Nubian** goats, with their distinctive Roman noses and long ears are attractive to many, although they are strong and can be difficult to handle. The earless **La Mancha** is popular in the USA but again, it needs considerable space and the males need firm handling. My own choice, if the goat is to be primarily a pet, would be the African Pygmy.

The great advantage of the African Pygmy is its smallness and the fact that it can be kept in comparatively confined areas. Like all goats it is a browser of weeds, but it will also eat a high proportion of grasses. It is cheap to feed, with one bale of hay lasting for about a month. It is an attractive little goat, normally found in shades of brown and black, but also with a lighter fawn coloration. Details of the African Pygmy are to be found on pages 21, 24.

Many goats kept as pets are not milked. Females will come into heat periodically during the breeding season, but if they are kept unmated, they will not produce kids and a flood of milk. Some females may produce milk without kidding. Such 'maiden milkers' will need to be milked to ease the pressure on the udder, but if a little milk is left behind each time, the amount will gradually diminish until she is 'dried off'. The African Pygmy produces comparatively little milk even when it has kidded.

Male or female?

A question which frequently occurs to those thinking of keeping goats as pets is whether to have males, females or both. Females will come into heat regularly in the breeding season. When this happens they will bellow for a mate for a day or two during each heat period and this can be wearing on the nerves. Close neighbours may also object, so it is an important aspect to consider before deciding whether to keep female pet goats.

Male goats may be preferable as pets because of this but, in my view, the choice should still be restricted to the African Pygmy breed, unless there is a particular interest in keeping strong males for harness work.

Males must, of course, be castrated when a few days old. If not, they will have an offensive smell and may be aggressive. It is advisable to have the scent glands on the head treated at the same time, so that there is no possibility of any smell developing. Castrating a female is not normally practicable. It is a major and expensive operation by comprison with male castration which is simple and straightforward.

Castrated males can be housed together. Similarly, a castrated male can accompany a female.

Better safe than sorry

All pet goats should be disbudded, as indeed should any goats which are regularly handled. Horns, even on a docile goat, are potentially dangerous. It is a simple procedure to have the horn buds treated at a few days old so that the horns do not grow. The procedure is explained on page 87.

If pet goats are being purchased for the first time, it is best to buy disbudded kids. If adults are bought, do ensure that they are hornless. I am aware that many people keep horned African Pygmy goats because they feel that the horns are an integral part of the overall appearance. Nevertheless, they are potentially dangerous and I repeat that families with young children should avoid horned goats.

The question is often asked – how safe is a goat? Individual goats will obviously differ in their temperaments, but as a general rule, most hornless female goats are harmless. Most castrated and hornless males are also harmless but the point has already been made that their strength

One of the author's British Alpine goatlings. If she had horns, they could present a hazard to the children.

may make them difficult to handle.

Goats are browsing, vegetarian animals and do not have the predatory instincts of a carnivorous animal such as a dog. They do not snap and bite, although they have a tendency to nibble. This may manifest itself in the nibbling of your sleeve or shoe-laces, but it is relatively harmless and goats will not bite in the way a dog may.

Newcomers to goats are sometimes nervous of what they see as the goat's tendency to butt. It is true that a goat will defend itself in this way if, for example, a dog should worry it, but it is a purely defensive tactic. As goats are herd animals, used to a herd leader, they will accept their owner in the role of leader, and will not butt him or her.

By the same token, the owner, having deprived the goats of their natural horned defences, must ensure that they are not left or tethered where dogs can worry them.

Housing pet goats

Pet goats are simple to house. A shed with an attached concrete run will cater for two animals. It is not humane to keep a solitary goat. They are gregarious herd animals, used to company. I have referred elsewhere to single goats which have befriended donkeys or other animals, and have lived contentedly with them. Owners of single ponies sometimes buy a castrated male goat so that he provides company for the pony. Again, this is a practice that can work well, as long as the relative temperaments of the animals are compatible. These are exceptions which disprove the rule but, as a general principle, a goat should be kept with one of its own kind. A male and female should not be kept together all the time unless the male has been castrated.

The type of housing sold for larger dogs is ideal. This is a shed with an attached run. A concrete run provides daily exercise and is easily cleaned by sweeping out and hosing down. Individual pens will not be required where only a small number of goats are kept, but there should be a well-placed feeder, drinker and hay rack. A mineral lick will also be appreciated. Full details of feeding and the different types of food needed by goats are given in chapter 7. Pygmy goats will obviously need less than the larger breeds.

A small house and run for pet African pygmy goats

The exercise run will need paving slabs or blocks so that the goats are able to climb up and down, leap and generally keep themselves happily occupied. They will need additional exercise in the form of daily walks, as would be the case with family dogs, and it is here that the value of lead training becomes apparent. A regular walk not only provides exercise, but enables the goats to browse for a proportion of their food. Country lanes, commonland, woodland and bridle paths all provide a range of foodstuff. The particularly popular wild plants, as well as the poisonous ones, are listed in chapter 7.

Training

It is never too early to start training and it is best to begin when the kids are a few days old. Get them used to wearing collars straight away and clip on a dog lead when taking them for a walk. The principles of training are exactly the same as those used for training a dog. The goat should learn to 'stay' and 'walk on' according to instruction. It should follow you, rather than dragging you along behind it. Remember that you are the herd leader. As with any form of teaching, firm and gentle persuasion is far more effective than ill-tempered bullying. A goat does not respond to punishment and will only be frightened by being smacked. Such treatment is likely to produce the opposite of that intended. Patience, by comparison, is invariably rewarded. A little bribery in the form of titbits will provide pleasant associations for the goat and make the learning process easier.

Grooming

All goats need regular care to keep them clean and healthy. The coat will require regular brushing and checking for the presence of parasites such as lice, mites and ticks. Where these are detected, a dusting of a proprietary louse powder from the vet should be applied. Pet goats and those competing in shows will tend to be groomed more frequently than those which are part of a commercial or small domestic herd.

The two areas which require particular attention are the coat and the feet. The coat should be brushed fairly regularly to remove any flaking skin and to promote an active blood circulation to the skin surface. Any matted areas of hair will be removed and the general condition of the coat will show a noticeable improvement with regular brushing. It is a good idea to have two brushes, one fairly stiff and coarse to remove tangles, and a fine soft one to impart a sheen to the coat. The feet need regular trimming to keep the nails from curling around the sole. Details of the foot trimming technique are given on pages 144-5.

Giving a goat a bath is not difficult, but it should take place only when weather conditions are appropriate. A tin bath which the animal can step into is ideal. A Pygmy goat can be bathed in a plastic baby bath. Ensure that the water is warm without being too hot and encourage the goat to step into it. If it is used to the procedure there should be no difficulty. For this reason, it is wise to accustom kids to being bathed from an early age. It is much more difficult to introduce the practice to an adult goat which is not used to being bathed. A mild baby shampoo or dog shampoo can be used, although purpose-made goat shampoos are now available. Avoid using detergents for these remove the natural oils from the hair and may cause allergic reactions. Swab the animal all over with a sponge dipped in the soapy water until it is quite clean, then rinse away the suds with clean, warm water. Rub the goat down well with a large towel and brush the hair. If it is a warm, sunny day, drying will take place rapidly. Do ensure that there is no wind, for it can cause chilling.

Browsers also make avid shoe-lace nibblers and pocket-riflers

A goat which is going to the show may require a few little extras in the form of beard trimming or clipping hairs which may be sticking out untidily. If the goat has had a bath, or if it is cold, the wearing of a rug is advisable.

The harness goat

Using goats in harness is an ancient practice. It is also popular with many modern goatkeepers. There are societies devoted to goat harness driving and their colourful parades have become a popular feature of goat shows in recent years.

Castrated males from the larger breeds are the best choice for harness driving, and the Anglo-Nubian or Nubian male is particularly suitable. It is a big, muscular animal of considerable strength. Training a goat to work with a harness is a skilled prac-

Goat in harness

tice and anyone interested in this aspect is recommended to join a goat driving society. Specialist suppliers sell goat harnesses, although many experienced drivers make their own.

Whichever breed or type of goat is chosen as a pet animal, it is important to remember that the scale of goat-keeping does not affect the quality of management. It is just as important to have a pet goat vaccinated against the range of clostridial diseases as the commercial goat. Similarly, the considerations of ruminant feeding which require a balance of concentrate feeds and roughage are just as vital to the pet animal as it is to the large herd. Finally, the owner of a single pet goat has the same legal requirement as the herd owner, to keep a record of all movements of the animal on and off the property. In the event of an outbreak of disease, this information may be required by the authorities in order to trace its source. Small may be beautiful but it is still related to the larger scale.

Daily constitutional for a small herd of goats

16. HEALTH

Prevention is better than cure. Despite being a hackneyed phrase, this is still a true saying. Many of the problems which affect goats have appeared because man has subjected them to the unnatural conditions of domestication.

The greater the degree of intensification, the more problems there are. Despite this, goats are remarkably free from problems, compared with other livestock, a reflection perhaps of their relative freedom from the excesses of agribusiness.

It is to be hoped that the newly flourishing Angora and Cashmere industry, with its embryo transplants and rounding up of the feral goat population, does not fall into the trap of extreme intensification that has bedevilled other fields of livestock farming. Intensive farming may have produced yields undreamt of by our ancestors, but it has also escalated health problems.

There are certain regular maintenance tasks to be undertaken where goats are concerned. They include foot trimming, vaccinating and worming, procedures which are necessary because of the restricted conditions we impose on them.

Apart from these their needs are simple; good dry housing, fresh air, clean land, a balanced diet of roughage and concentrates, clean fresh water, common sense and affection. A regard for these will keep goats healthy, happy and productive, with few problems. Nevertheless, it pays to be prepared for trouble.

A hospital pen

Having somewhere to put a goat if she is off-colour is a good idea. She has peace and quiet in which to recover and convalesce. In the event of there being an infectious condition, the other goats are protected. The pen should be in the quietest part of the house, away from the hub of activity, but where she can still see the other goats at a distance. Goats are gregarious and are distressed if they are on their own.

Sheep hurdles are excellent for making a temporary hospital pen. They slot together quickly and easily, although the gaps in the walls are draughty. The best solution here is to place straw bales outside the pen to provide warm, well insulated walls. If the pen is equipped with a heat lamp, it can be used for all sorts of emergencies such as kiddings in particularly cold weather.

The first aid kit

Being prepared for all eventualities includes having a first aid kit to hand in the goat house. Most of the contents are available from three sources: the chemist's shop, the local veterinary surgeon and the specialist supplier of goat equipment.

Clinical thermometer
Scissors
Sterile cotton wool
Sterile bandages
Antiseptic cream

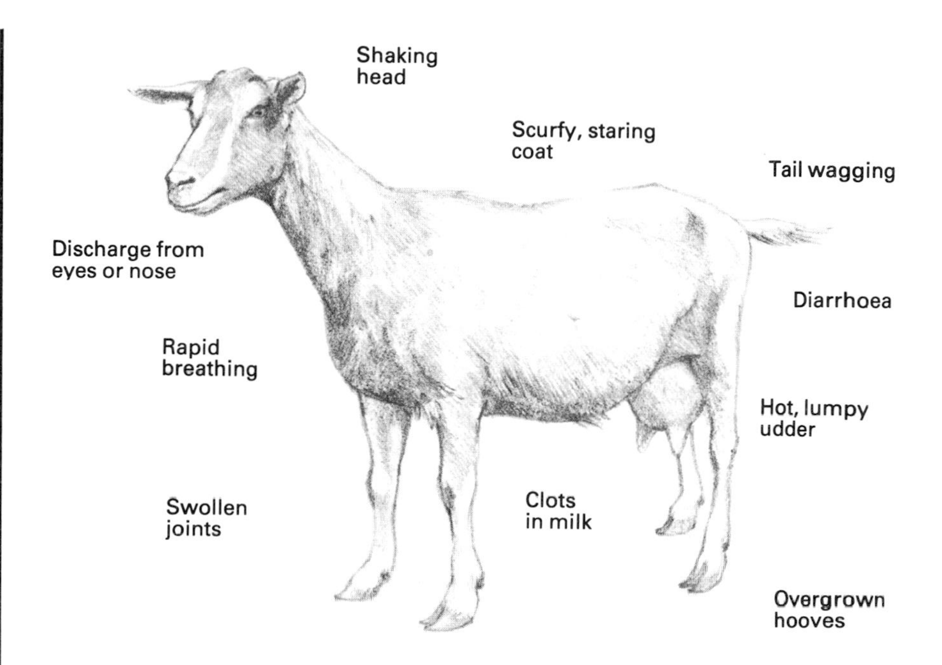

Signs of ill health to watch for

Disinfectant
Plastic drenching bottle
Tweezers
Hoof paring knife
Surform plane
Iodine
Tetramycin or Aureomycin
spray
Glucose
Kaolin solution
Cooking oil
Udder cream
Tin of molasses or treacle
Vaseline

In addition to these basic commodities will be the preparations for worming and other veterinary prescriptions which will come from the vet as needed.

Before the vet comes

When it comes to diagnosis of animal ailments, the vet is the only one who can be definite. Although experienced goatkeepers may often be correct in their diagnoses, there can be times when they are disastrously wrong. Symptoms can vary widely from one goat to another, and different conditions may have confusingly similar characteristics.

The golden rule is that if there is any doubt, call the vet. He or she may be able to offer advice and reassurance over the telephone and will be able to judge whether a visit is necessary. In that respect, it is helpful to be able to give a clear description of the symptoms as you see them, and to indicate what the goat's temperature is.

It is not difficult to take the temperature, as long as you establish which is the right end to test. It is no use popping it under the goat's tongue; not if you wish to see it again. No, I am afraid that it is the other end!

If you lift up the tail slightly, you can pop in the thermometer without her noticing if you are discreet. It only needs to be there for a minute before removal. The normal temperature of a goat is around 39-40°C (102-104°F), while the average pulse rate, at rest, is between 75–90 beats per minute.

Armed with these facts, as well as a description of the symptoms, you are in a good position to help the vet as much as possible. In the event of an emergency or possibly a life or death situation, all this is academic. The thing to do is get on the telephone and get him to come immediately.

Such an occasion is no time to discover that you cannot find the vet's number. It should be pinned up in a prominent place.

Foot trimming

Goat's feet need regular trimming. Since they left the rocky heights of feral life, which would have kept their hooves short, they have to rely on the regular administrations of humans.

The easiest way of doing it is to tie the goat by the collar, or yoke her to a stand, so that she is not able to move too far in any direction. Giving her some tit-bits in a bucket, or an armful of greens, is an excellent way of keeping her occupied.

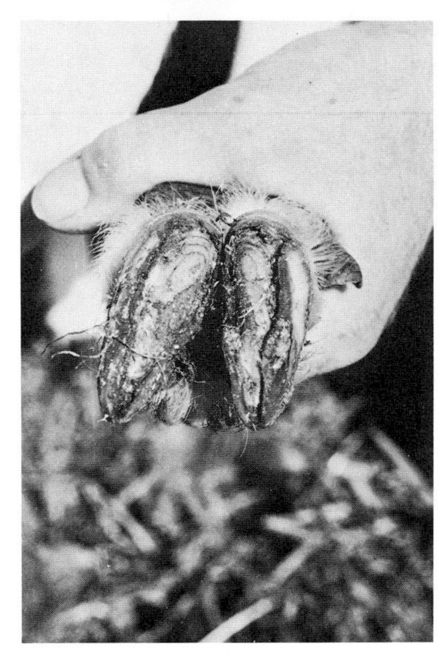

These hooves need trimming!

A sharp knife and a Surform plane are the best tools for the job. The knife does the main paring, while the plane smoothes it off. Lift the first foot gently, but with a firm hold, and bend the leg in its natural position backwards, without taking it too far back to cause discomfort to the goat.

She will probably try to kick her foot free, but if you keep a firm grip and reassure her gently, she will eventually desist, especially as her attention is on eating. Goats are single-minded, and a tit-bit in the bucket is worth any number of feet on the ground.

The aim of the trimming is to cut back the sections of nail which have grown and possibly curled under the sole. Pare these back carefully, doing a little at a time so as not to cut into the

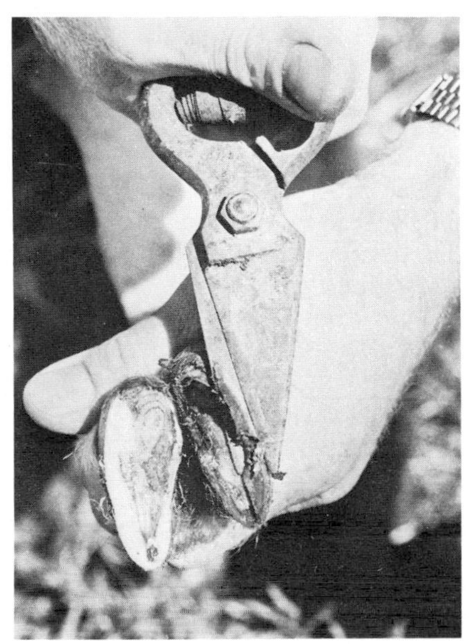

Hoof trimming: having successfully pared back the left side, the handler is now tackling the excess growth on the right

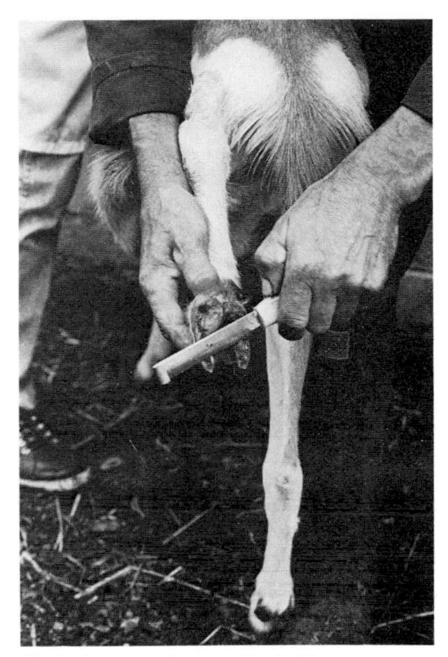

Hoof trimming using a paring knife

sensitive areas. Clear out any soil or bits of gravel lodged in the hoof, then use the plane to smooth off.

The Surform plane is much more effective than a normal plane or file. Its action is like that of a grater. If it is used to file the goats' feet fairly regularly, it should not be necessary to use a knife because the hooves will never get to an overgrown state.

Once the pedicure is over, give the goat a lot of praise for her cooperation. She will be even better behaved next time, as long as there is something worth having in that bucket.

Worming

Any goat which is out on pasture is at risk from internal parasitic worms. The most common are **roundworms** which can infest the gut. They lay eggs which are shed in the droppings and contaminate the herbage. If goats eat the contaminated herbage they are eating roundworm eggs, and so the cycle starts again. Sources of danger of roundworm infection are:

- pasture where goats or sheep have grazed within the past year
- paddocks or areas of strip grazing which are used in rotation
- herbage cut from pasture which has been spread with goat manure

Safe sources of browse are:

- newly sown pastures
- pasture grazed by cattle or horses, but not goats or sheep
- land where hay has just been cut

– browse material such as hay, fodder crops and cut branches

Goats can tolerate a certain level of roundworm burden, and they develop some immunity. Kids and young goats are at the most risk, as well as stall-fed goats that are suddenly introduced to pasture. They have no immunity and will suffer more than those that have acquired resistance over a period of time. Symptoms of severe infestation are loss of weight, reduced milk yields and a harsh, staring coat.

Other types of worm There are three of these:

(1) **Lungworms** affect the lungs and air passages, causing frequent coughing and a more rapid breathing rate. The intermediate stage of this parasite is carried by snails and slugs so it is more prevalent in wet pastures.

(2) **Liver fluke** occurs in the same conditions. The best way of avoiding both these pests is to keep goats off wet pasture. If necessary it can be drained.

The Ministry of Agriculture issues a 'fluke forecast' for the benefit of sheep farmers in Britain, rather like the Meteorological Office's 'pollen count' for hay fever sufferers. If the forecast is for more flukes than normal for a particular period, it is best to heed it and keep goats off the pasture.

(3) Goats can be affected by **tapeworms** which are carried by dogs. It is in everyone's interests to keep dogs away from pasture and areas where goats are likely to be grazing.

When to give a worming dose All goats which are on pasture need regular worming, unless the area available to them is so vast that they are unlikely to be on the same ground more than once a year. The time to administer worming preparations is a few days after kidding, in the case of the mothers, and at the age of about five months for the kids.

A second dose in autumn, before they are winter-housed, will ensure that they continue relatively free.

If you are making intensive use of paddocks, more frequent dosing such as once every six weeks may be necessary in the summer.

The best way to decide on this is to consult the vet who will, if necessary, take a 'worm count' from a sample dropping in order to assess the situation and advise on appropriate action.

Worming procedure A wide range of **anthelmintics** (worm killers) is

Drenching a goat, using a syringe-type applicator

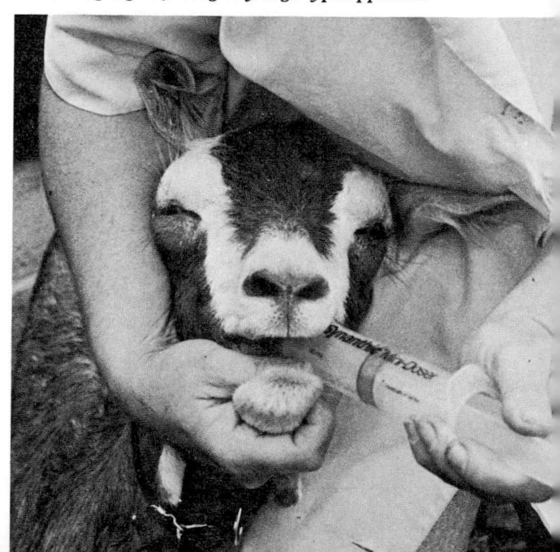

available. They include **thiabendazole**, **fenbendazole**, **albendazole** and **oxfendazole** which are available under a variety of trade names. The most common ways of dosing goats are **(1)** a **drench** (dose of liquid medicine given to an animal by mouth), **(2) worming tablets** and **(3)** a paste which many goatkeepers claim is the easiest method.

The drench is given in a plastic bottle with a curved tube which is inserted in the side of the goat's mouth, in the gap between the front and back teeth. It can be rather a messy procedure. The key to success is to hold her securely, tip her head back slightly and not miss. A syringe applicator with a long tube helps.

I have always found tablets impossible to administer. Many goatkeepers claim that there is nothing to it, but if there is a secret, I have never discovered it. The tablet would be popped in the mouth, reassuring words would be spoken and she would obligingly swallow. Flushed with success, I would release the goat and praise her – only to see, a few moments later, a white tablet flicked casually out of the side of the mouth.

Vaccination

Every goat should be vaccinated once a year against the range of clostridial diseases which include enterotoxaemia and tetanus. The procedure is a simple one and well worth it. The vet will advise you and demonstrate how the injection is to be given.

AILMENTS A–Z

Abortion There are many reasons why a goat should abort, from accident and stress to deficiency disease and infection. The vet should be called without delay.

Abscess A bacterial infection forming pus under the skin and usually the result of an infected wound. Do not lance but apply a poultice over a period of a few days. A simple one is to take a little Vaseline and mix with it the same amount of honey, then apply the mixture to the abscess. The honey exerts an osmotic pressure on the pus, drawing it to the surface and making it come to a head before bursting.

Once it has burst it can be cleaned out and antiseptic cream applied. If it has not come to head in three days, it may be a **cyst**, a more serious type of swelling for which you will have to consult the vet.

The treatment is effective for an ordinary abscess, but if it is a cyst it will do no harm.

Acetonaemia (Ketosis) This is essentially a chronic shortage of carbohydrates in the diet. The goat may be pregnant with large kids or a heavy milk yielder not being fed enough to cater for her needs. Ketones are formed in the blood and there is a smell of acetone or nail varnish on the breath. Milk yield declines and there may be off-taints in the milk.

Put a little molasses or black treacle in the drinking water and warm it up to make it more attractive. As a fast-acting treatment give a glucose and warm water drench. As soon as she begins to show more interest in food and in the world generally, increase her concentrate cereal ration and give good quality hay. The vet should be called, particularly in the case of a pregnant goat.

Anthrax The extremely dangerous Anthrax is a notifiable disease. In other words, its presence, or suspected presence, must be reported to the Ministry of Agriculture or the police. It is extremely rare in Britain, thanks to the strict regulation which requires that any sudden, unexplained death of livestock is reported. There is no treatment and death is rapid. It can affect other livestock and humans and remains active in the soil for many years.

Bleating Continuous bleating can be expected on the following occasions: (1) when the female is on heat; (2) when she has been separated from her normal companions; (3) when she is out in the rain (possibly tethered) and cannot get to shelter. On any other occasion, continuous bleating can be taken to be a sign of distress. If there is no apparent cause and it continues, call the vet.

Bloat The rumen of the goat becomes blown up with gases which are unable to escape. Too much lush spring grass can cause it, or an excess of a particular food which disturbs the balance of the rumen's workings.

The goat's left side bulges out and she may bleat or pant in distress. A drench of cooking oil will help to get things moving, and massage may help, but these courses of action are where the condition is not too far advanced.

If she is like a balloon and in an obviously life-threatening situation, the vet should be called immediately. He may puncture the rumen high up on the left side in order to release the gases, not something to be attempted by the layman.

Always ensure that goats have hay to eat before they go out to spring pasture, and restrict the time spent there in the first weeks.

Blood in milk See PINK MILK.

Brucellosis There are two kinds of brucellosis, *B. melitensis* which affects goats in many different part of the world, but is unknown in Britain, and *B. abortus* which affects mainly cattle but can also affect goats. A rigorous control programme in relation to cattle has now eliminated the disease in Britain. Both forms of the disease can cause undulant fever in humans if infected milk is drunk. Pasteurization of goat's milk is essential in those areas of the world where brucellosis is still to be found.

Caprine Arthritis Encephalitis (CAE) A comparatively recent introduction to Britain, CAE has been apparent in the USA and other parts of the world for a considerable time. It belongs to the family of viruses known as **retroviruses**, a grouping which also includes HTLV III or the AIDS virus.

This has raised fears amongst many goatkeepers that there is a risk to humans from keeping goats which may be carriers. However, there is no evidence to indicate that the CAE virus affects humans, or that the AIDS virus affects goats, for they belong to different subfamilies.

CAE in relation to goats must be taken seriously for there is currently no cure or preventive vaccine. Although it is still comparatively rare in Britain, every goatkeeper must take measures to prevent its spread. Goats can be positive carriers of the slow-acting virus, but the full disease may not necessarily develop. Its presence can be detected by the presence of antibodies in the blood.

Symptoms are lack of coordination, swollen joints and paralysis. It appears to be passed on to the kids via the colostrum, with young kids developing an encephalitic form of the disease, while older ones show an arthritic form.

If a kid is taken immediately from its positive carrier mother and colostrum given from another goat, it can be kept free of the virus. There is evidence that it can be passed on at goat shows, where kids may be fed milk from infected sources.

Males have also been infected by positive does during mating, although it is claimed that a positive buck will not necessarily pass it on to a negative doe. However, since the virus can also be transmitted through cuts in the skin, and these could occur any time a doe is brought to stud, for all practical purposes the risk to females from positive males is the same.

There is really only one long-term approach and that is to have all goats blood-tested and unless they can be isolated, have the carriers put down.

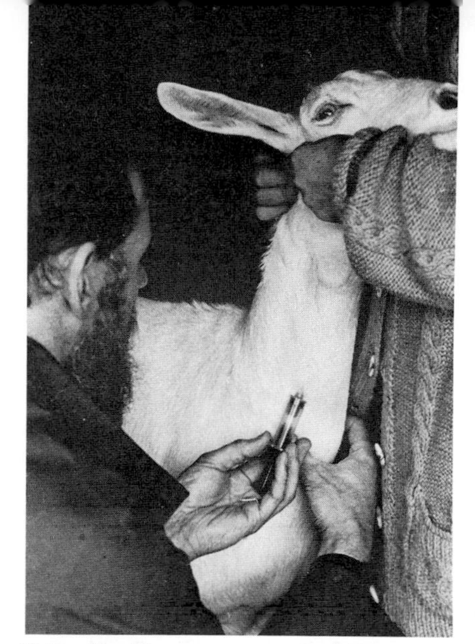

Taking blood from a goat to test for CAE

Do not allow other people's goats to come on to your site unless they are known to be blood-tested and clear.

Many goatkeepers offer boarding facilities for goats when their owners are on holiday, just as many people make use of such facilities for their own animals. It is in the interests of both sides that they deal only with clear stock.

Similarly, goats taken for mating should be clear. Many stud-owners now insist on a current blood-test certificate before they will accept a goat for service. Owners of does should insist on similar confirmation with regard to the buck, and anyone considering buying goats should ensure that they come from a CAE-free herd.

The British Goat Society and the Ministry of Agriculture operate schemes of testing and monitoring which will enable goat owners to apply for CAE-free herd status.

Caesarian birth In the event of a difficult kidding and inability of the kids to be born normally, the vet will probably decide on opening the abdomen wall to save them. Any goat which is obviously straining without success for more than half an hour needs urgent veterinary attention.

Circling disease (Listeriosis) The microorganism *Listeria monocytogenes* affects the brain, making the goat lose coordination and stagger around in circles. It may also cause blood poisoning and abortion. Other symptoms are high temperature and pressing the head against the walls of the pen.

The organism is soil-borne and is found in poorly-made silage. As comparatively few goats in Britain are fed on silage it is rare. Where silage is fed, great care should be taken to ensure that it is uncontaminated.

It is really not feasible to make silage on a small scale for the problems of effective exclusion of air from the fermenting grass are considerable. On a larger scale, it is wise to have samples of silage tested for quality by the Ministry of Agriculture.

Cloudburst (False pregnancy) The goat is mated and shows all the symptoms of being pregnant, but what emerges from the womb is a burst of liquid. It is relatively common in Britain where many goats are left unmated until their second year. Apart from the inconvenience of missing for that year, the cloudburst female usually comes to no harm and kids normally the following year.

Coccidiosis This disease is caused by single-celled parasites called **coccidia** which are found in warm, damp areas such as the bedding of housed goats. A certain tolerance is shown as long as the burden does not become too great.

Coccidia are passed out in the droppings and are subsequently ingested, causing reinfestation. Kids are particularly at risk, showing symptoms of diarrhoea, sometimes with blood in the droppings, and general straining and stress. The vet should be called and the usual treatment is a drench for individual cases and general dosing in the drinking water for the rest.

With proper management and a humane approach to livestock keeping which resists the temptation to intensive stocking, this condition should never appear. Coccidiosis which affects poultry does not affect goats, but that found in sheep does. For this reason, goats and sheep should not be kept together.

Colic Overfeeding concentrates, suddenly changing the diet, or giving too much of any one foodstuff can all cause digestive upsets. This is not serious in the way that bloat is, but it can be painful and distressing to the goat. The best thing is to walk her up and down, massaging the rumen. It does not normally last long.

Collapse This is a general name to describe the goat which collapses and is unable to move. It is important not to leave her on her side in case her breathing becomes restricted. Prop her on her front with a straw bale on each side and put an old blanket over her. Call the vet immediately. It could be any number of things but expert help is needed without delay.

Cough Goats do cough, and sometimes they sound just like humans when they do it. Most of the time a cough is nothing to worry about, but check that the concentrate feed is a nice coarse ration without too much powder in it. Fine dust in a feed can cause a cough, as well as a sneezing bout. More serious is the persistent cough, particularly if there is a high temperature. In these cases, consult a vet.

Cuts See WOUNDS.

Diarrhoea There are a number of conditions which can cause diarrhoea, from a simple digestive upset to a more serious bacterial infection. As an immediate treatment restrict all foods except hay and give warm drinking water with glucose in it. A kid could be given this in a feeding bottle.

A kaolin preparation is useful to help stabilize the droppings. If the scouring does not clear up by the following day, or if the droppings become bloodstained, call the vet, for it is more serious.

Enterotoxaemia No goat should ever have to suffer this condition, which is the worst of the goat diseases, because to vaccinate against it is so simple and effective. It is one of a family of soil-living organisms which cause tetanus, pulpy kidney and other serious but avoidable conditions.

Symptoms are a drunken appearance and ultimate collapse of the goat which has severe diarrhoea and pain. She will often lie on her side, paddling her legs pathetically. Anyone who believes that vaccinations are 'unnatural' should see a case of enterotoxaemia.

Fleas Yes, even pedigree goats get these! A proprietary treatment such as that used for dogs and cats is effective. The bedding should be changed frequently to avoid a build-up of them.

Fly strike A common problem with sheep, fly strike is not common with goats. If areas of the body become soiled with droppings, as in a case of scouring, or if there is an open wound, bluebottle flies will lay their eggs there. In a short time these will hatch to form a boiling nest of maggots that tunnel into the flesh.

Short-haired goats are rarely in this condition, whereas sheep, with their concealing wool, may have such an infestation under the surface. Angora and Cashmere goats are obviously more at risk because of their coats.

Any goat which is seen to be pawing the ground or rubbing against a fence in the summer months should be investigated for fly strike. If it is discovered, clip back the hair, clean out the maggots and pus with disinfectant and apply antiseptic cream. Then spray the surrounding area with fly spray.

Flies will occasionally lay their eggs in the nostrils so that the tunnelling maggots can cause considerable damage, but this is not as common in Britain as it is in hotter climates.

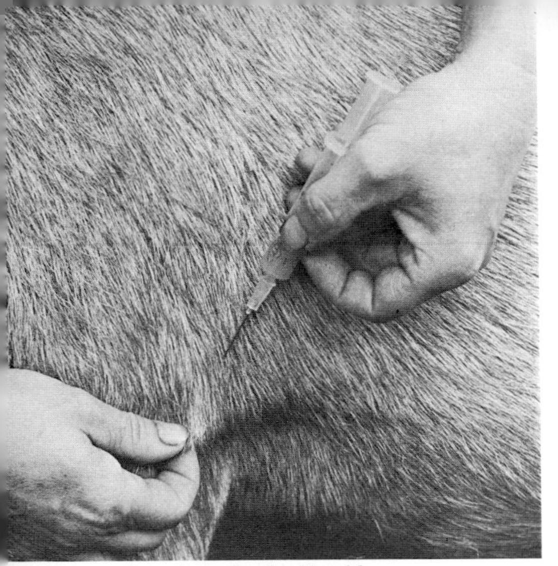

Injecting under the skin with enterotoxaemia vaccine

Foot and mouth disease This is a notifiable disease which can affect all cloven-hoofed animals. It is rigorously controlled in Britain, where it is normally absent as a result. When it does occasionally flare up, the Ministry of Agriculture's policy is to slaughter all affected livestock and quarantine the area.

Symptoms are lameness and dribbling at the mouth, which often has blisters around it. If these combined symptoms appear, you must notify the vet, Ministry of Agriculture or local police.

Foot rot Again, this is more commonly seen in sheep, but can occur in goats if they are on marshy ground in summer, or spend much of their housed time on damp, dirty floor litter. The bacterium *Fusiformis nodosus* enters the softened area of the hoof and sets up a local infection. This has a vile smell and quickly leads to lameness.

The goat should be transferred to a concrete surface and separated from the others. Trim back the hooves and scrape out the areas of infection, cleaning the feet with soapy water and disinfectant. Dry them then confine her to a concrete area until it heals up.

If you suspect that the other goats may have been in contact with it, but there is no evidence, it is a good idea to take preventive action to stop the disease spreading. Get the goats to step into a container of disinfectant, then stop them going on to the area where they have been grazing for at least a week. The organism is a fairly short-lived one and will not survive for long on the pasture.

Goat pox Groups of small, raised watery spots are occasionally found on the udder and around the mouth. They eventually form blisters, break and form crusts which then fall off. Treat with an antiseptic cream and be sure to milk this goat last so that there is less risk of transferring the infection to the others. Be sure to wash your hands well after handling her.

If the condition does not clear up after the crusts have fallen off, consult the vet. It is not unknown for laymen to confuse this relatively mild condition with the more serious ORF.

There is a malignant form of this condition in Africa and Asia but it is unknown in Britain.

Goitre The thyroid gland in the throat swells up as a result of iodine deficiency. As the thyroid gland has a major influence on the metabolic rate of the body, it is a potentially serious condition, but reversible.

If goats are given a mineral and vitamin supplement in their concentrate ration, this condition, or indeed any other deficiency trouble, should not occur. Care should be taken not to feed too much kale or other members of the brassica family such as cabbage and turnips. They have the effect of 'locking up' the available iodine if eaten to excess.

The vet should be called if the effect of a normal feed supplement does not produce a noticeable improvement. It may be necessary to give larger doses of iodine but as it is a toxic substance, it should be left to the vet to decide what a safe dosage would be in the circumstances.

Grass staggers/Grass tetany See HYPOMAGNESAEMIA.

Heat stress Even in Britain it occasionally gets so hot that goats suffer. Although they can stand a lot of heat, they should never be tethered in hot sun without access to shade, shelter and water. It is such an obvious thing to write, and yet there are still people who do it, like those who go to agricultural shows leaving their dogs locked up in cars to bake.

White-skinned goats such as Saanens can suffer from sunburn. There are also incidences of skin cancers on Saanen goats in Australia.

Husk The symptoms are frequent coughing caused by parasitic lungworms which, in turn, produce a form of bronchitis. Consult the vet who will prescribe the appropriate wormer. All goats should be wormed if they are on pasture.

Hypocalcaemia See MILK FEVER.

Hypomagnesaemia Grass staggers is a descriptive name for this condition, because it frequently arises when animals go out on to new spring grass, and the main symptoms are lack of coordination and staggering.

It is caused by a deficiency of magnesium when the flow of milk production depletes the body reserves. The vet should be called immediately to give a magnesium injection. If left untreated, the condition quickly becomes fatal.

Milking goats lose a considerable amount of minerals in their milk output which is far greater than that of cows in relation to their size. A mineral and vitamin supplement as part of the concentrate ration will ensure that such deficiencies do not occur. Goats should also be given hay before they go out on new pasture in spring, and their time spent there should be restricted at first.

Johne's disease Caused by the bacterium *Mycobacterium paratuberculosis johnei,* this is a condition affecting cattle, although in rare cases it affects goats. Infected animals suffer from a tuberculous condition of the intestine which ceases to be able to absorb nutrients. As a result, weight is lost rapidly and the animal weakens and dies.

There is no treatment and the condition is invariably fatal. Infected animals should be slaughtered and no livestock kept on the site for at least two years; the organism can survive on pasture.

Although rare, this disease should

be taken seriously by all goatkeepers. Anyone buying goats should obtain assurance from the vendor that there have been no cases of the disease on the site. Early diagnosis can be made from a laboratory examination of the droppings.

Joint ill See NAVEL ILL.

Laminitis This is lameness brought about by an inflammation of the hoof tissues. It is usually the result of an imbalance in feeding rations, as for example when too much spring grass is eaten. Too much concentrate in relation to hay can also contribute to it. Consult the vet who will prescribe treatment to reduce inflammation.

The long-term treatment is not a reliance on drugs but a realization that goats can not be forced into excessive production by overfeeding of rich foods.

Lice Any goat can pick up lice and it is important to keep a watch out for them. They are small and greyish and are found at the base of the hairs. A proprietary lice or flea powder such as that used for dogs is suitable. Change the bedding more frequently.

Listeriosis See CIRCLING DISEASE.

Louping ill This is a virus infection causing encephalitis (inflammation of the brain). It is transmitted by infected ticks and is found primarily in sheep. It is claimed that only goats in the mountainous areas are likely to be infected because this is where the ticks are found. As ticks, sheep and goats are found in lowland areas as well, the logic of this claim is puzzling.

Liver fluke A parasite of wet areas with its secondary host as the snail. See WORMING.

Maiden milker It can sometimes happen to males too, even stud billies. Again the milk should be drawn off, leaving a little behind each time, until it ceases of its own accord.

Mange See MITES.

Mastitis I have said in the section on Milking (page 98) that it is important to check the milk for the presence of clots which could indicate mastitis. The condition is an infection and inflammation of the udder, and varies from very mild cases to massive and chronic infection.

A mild case will often cure itself if the udder is given a chance to do so. An immunity will develop, as long as the udder is massaged to increase the blood flow, while avoiding the stripping out of all available milk.

The trouble arises when the animal is regarded purely as a milk machine. At the first sign of trouble, antibiotics are given which may knock out the infection, but do not allow the body's defences to come into operation. There is then no natural immunity.

Next time an infection comes it will be a little worse, more antibiotics are used, and so the escalation continues. The worst cases now are where antibiotics no longer work, the infection is chronic, in some cases leading to gangrene of the udder.

I am not suggesting that there is no place for antibiotics. They are vital, but as last-resort strong drugs they should be kept for serious situations where an infection has overcome the body's defences.

If a mild case of mastitis develops in a young goat, the sensible thing is to take her out of family or commercial milk production and give her a chance to fight it herself. She will still need to be milked of course, but the milk should be discarded.

Meanwhile, try the traditional approach of massaging the udder with a compound which produces warmth in the udder area. This has the effect of increasing the capillary blood flow and bringing more white blood cells to fight the infective cells. As they fight, antibodies capable of resisting future invasion are produced.

A recently introduced liniment for udder massage is a product based on Japanese Peppermint oil. The liniment is applied at the first sign of trouble and creates warmth in the tissues. (It has been used for rheumatism for the same reason.) Several dairy farmers have claimed that this preparation (**Uddermint**) has cleared up mastitis and udder sores which were previously resistant to treatment.

Where heavy mastitis infection is evident, there may be no alternative but to use antibiotics, and the vet's advice should be sought. It is also vital to have a laboratory examination of a milk sample to establish which strains of bacteria are causing the trouble. It sometimes happens that an unsuitable antibiotic is prescribed.

To take a sample, use a strip cup

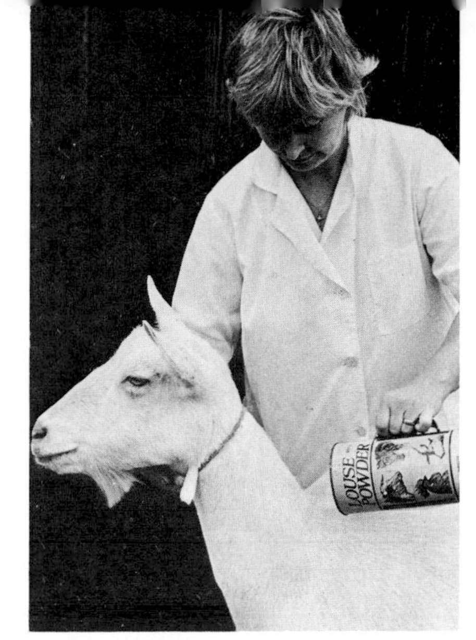

Louse-powdering a goat

with a black removable inner dish; this allows close examination of the milk so that you can spot any clots without difficulty.

When mastitis is treated with an antibiotic, an intermammary syringe is used. This is like an ordinary hypodermic syringe except that it is furnished with a plastic tube at the end, not a needle. After milking, the affected teat is held near the end, and the end of the tube inserted into the teat opening. Gently slide the tube up the teat canal and press the plunger to empty the contents inside.

The course of treatment continues for several days. During this time the goat is milked in the normal way, but the milk from the affected quarter is discarded. Once the infection has cleared, a minimum of three full days must elapse after the last treatment before the milk is safe for consumption. Before that, there will be antibiotic residues in the milk.

Maedi-Visna (MV) Mainly associated with sheep, the MV virus can also affect goats. Symptoms are nervous behaviour and gradual loss of condition, with eventual wasting away. It can be detected by blood tests. The Ministry of Agriculture operates a voluntary scheme whereby sheep and goat owners can have their animals blood-tested for MV-free status.

Metritis This is an infection of the womb which can be caused by introducing the hand into the womb to help during kidding, or where all or part of the placenta has been retained. It is recognized by a brownish, nasty-smelling discharge from the vulva. (A lightish coloured discharge is normal for a few days after kidding.)

The goat is off her food and there may be a higher than normal temperature. The vet should be called and he will normally prescribe antibiotic treatment.

Udder showing the drastic results of a bout of mastitis

Milk fever (Hypocalcaemia) A sudden deficiency of calcium in the blood, where rapid milk production drains away the body's reserves, can cause collapse of the goat. It usually occurs in the period immediately after kidding. Death will follow collapse unless a calcium injection is given without delay. Call the vet as a matter of extreme urgency.

He will normally give an injection of calcium borogluconate. Recovery is usually magical: the goat gets to her feet as if nothing untoward had happened.

Avoid feeding too high a level of concentrates in the period immediately before and after kidding. Do not milk her out completely for the first week, to avoid putting too much strain on her. Feed good quality hay at all times.

Mites There are many different mites, some on the surface of the skin, some burrowing into the skin and some getting into the ears. They can all be treated with a normal insecticide such as that used for dogs, but the burrowing one is not as easy to detect. If there is an obvious skin condition which is causing discomfort, consult the vet.

Ear mites should be suspected when a goat shakes her head often. A mite preparation can be sprayed into the ears to kill them. It is better to do it twice, with a couple of days in between applications. This will ensure that any eggs which subsequently hatch are also killed.

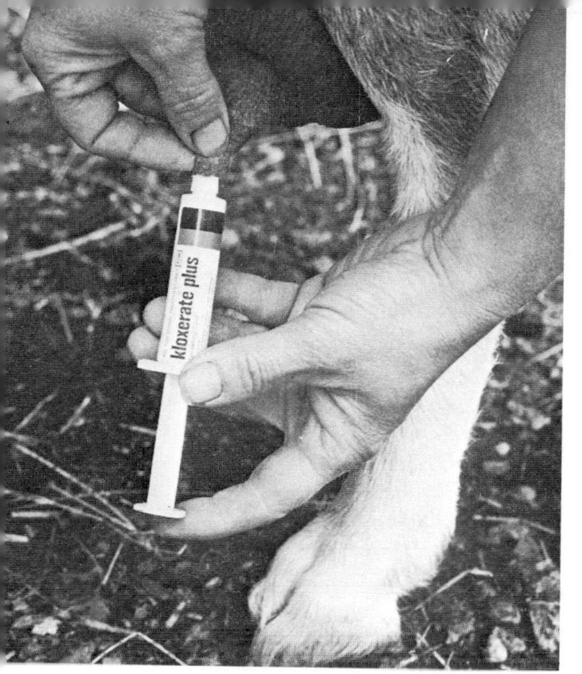

Inserting a tube of antibiotic into a teat canal to cure mastitis

Mange mites are some of the most troublesome of the burrowing mites, often producing secondary infections and sores. Veterinary examination and treatment is essential because there are several different types of mange conditions.

Navel ill (Joint ill) Found in kids, the cause of this bacterial infection is almost invariably a dirty environment. Infection enters the severed umbilical cord, producing a red swelling around the navel. The infection can travel to the limbs, producing swollen joints. The vet will prescribe a course of antibiotic treatment, but once the infection has affected the joints, recovery is unlikely.

Kidding pens should be thoroughly cleaned and disinfected before use, and clean fresh straw used as bedding. At the birth of the kids each one should have a squirt of aerosol based antibiotic such as aureomycin or tetramycin on the severed umbilical cord. An alternative is to dip the end of the cord in iodine solution.

New Forest eye See PINK EYE.

Orf (Contagious pustular dermatitis) This is a highly contagious virus infection which can be caught by humans. Nasty blisters form around the corner of the mouth and nose. Isolate the goat and call the vet immediately. Wash your hands thoroughly after handling the animal and do not let children go anywhere near. Occasionally there is a rare form of malignant Orf.

Pine (Fading disease) Pine is an apt name for this condition. The animal appears to pine away for no reason. But there is a substantial reason. It is caused by a deficiency of the trace element cobalt in the diet. It is this element which enables the body to produce vitamin B_{12}. It is more common in those areas where there are low levels of cobalt in the soil, and this is reflected in the quality of pasture.

The immediate treatment is for the vet to prescribe cobalt sulphate, but a more long-term solution is to give a cobalt 'bullet'. This is swallowed and gives off cobalt slowly over a period of months. A good mineral and trace element supplement formulated for goats, such as **Caprivite**, will ensure that sufficient levels of cobalt and other minerals and trace elements are present in the diet.

Pink eye (Contagious opthalmia) (New Forest eye) This is a contagious eye infection which affects the eye membranes, making the eye red and sore. The eye weeps and there may be pus. An antibiotic eye ointment applied in the inner corner of the eyes will clear it up. Keep the goat isolated and wash your hands thoroughly after handling her.

Pink milk A new milker or a heavy yielder will occasionally produce milk which is slightly pink. If you put it in a bottle and look at the bottom you will see the blood collecting there in a layer. It is the result of a few blood capillaries breaking in the udder and is normally nothing to worry about. The milk should not be sold because of its appearance.

The condition usually clears up of its own accord within a couple of days. If it does not, consult the vet in case it is something more serious.

Pneumonia There are several forms of pneumonia and any persistent cough should be investigated by the vet. It may be causd by lungworm (see HUSK) or it could be a bacterial infection which can produce a high temperature and serious lung damage.

Veterinary treatment is essential, and careful nursing is almost as important. Goats are not good patients. They have a tendency to give up too easily. The relationship between the goat and her owner is important here, and can often make the difference between recovery and decline.

Make a fuss of a sick goat. Give her a blanket to keep her warm, particularly if she has an infection like pneumonia, and talk to her frequently. If there is a good relationship, she will respond, and this in itself can help her back on the road to recovery.

Poisoning Goats have an insatiable tendency to nibble and should be kept well away from areas where known poisons are present. Examples are paintwork with lead-based paints, timber newly treated with wood preservative and areas where there is machine oil or diesel oil.

The chances of such poisonings are rare: more common are cases of poisoning by toxic plants. While many plants are toxic if eaten to excess, comparatively few are so poisonous that a relatively small amount will kill. Goats appear to have an ability to eat a certain quantity of toxic plants with impunity, but this does not mean that they will necessarily survive where other animals would die.

It is an area where much research remains to be done. A good coverage can be found in *Poisonous Plants in Britain and their effects on Animals and Man*, published by the Ministry of Agriculture. It is a book which should be on every livestock keeper's bookshelf.

Pregnancy toxaemia See ACETONAEMIA which is virtually identical.

Premature birth Occasionally seen in goats, the kids have a chance if they can be made to breathe, can be kept warm and can take in colostrum. Clear the nostrils of mucous membranes and rub them all over with a towel until they breathe. If rejected by the mother keep them in a box by the fire. Feed warmed colostrum and glucose with an eye dropper every hour. Once they are able to suck of their own accord, they will probably survive.

Prolapse This may be one of two types; the emergence of the cervix and vagina, seen as a mass of red tissue at the vulva, or the complete emergence of the uterus after kidding. Both conditions require the urgent attention of the vet who will replace them, using sutures and antibiotic treatment. It is not a task to be attempted by the layman.

Ringworm (Dermatomycoses) A fungus infection, it is seen as a circular red patch with flaking skin and loss of hair. It can be transmitted to other goats and to humans and should be treated with care. The vet will provide a fungicide which will kill off the organism.

Roundworms See pages 145-7.

Scurf This is a general name to describe loose, flaking skin. It is frequently seen in goats which have been winter-housed. It normally improves once they are able to get out into the sunshine.

A friend of mine claims that her goats used to stand in the warm spring rain if they had scurf and that the combination of this and the sunshine cleared up the condition.

It helps to give them a good comb or brush to remove the flaked areas. In warm weather a good wash and a shampoo followed by a brisk rub-down with a towel works wonders. Dog shampoos or baby shampoos which do not irritate the eyes are ideal.

Ensure that they have a mineral, trace element and vitamin supplement in the concentrate ration. A deficiency of zinc can contribute to scurf.

Stones (Urinary calculi) Male goats are prone to this condition where the urine flow is restricted by a blockage. Veterinary advice should be sought for it can cause extreme pain.

Hard water areas can contribute to the problem, and the use of rain water or a fitted water softener should be considered. It is important that males are not fed too much concentrate in relation to roughage. Dried sugar beet should not be given except in small quantities, and only then if it has been well soaked first. The condition can also be aggravated if he does not drink enough water. Encourage him to drink as much as possible by giving warm water.

Swayback (Enzootic ataxia) This is a condition which is rare in goats, being usually associated with sheep. It occurs in kids whose dams have had insufficient copper in the diet. The result is poor neural development so that coordination is incomplete. When the kid tries to get up it sways from side to side at the back.

There is little that can be done for the kid and it is better put down. Giving the mother a mineral and trace element supplement will give protection against this condition in future.

Tapeworms See Worming section on pages 145-7.

Tetanus (Lockjaw) A soil-borne organism which causes paralysis. There is no excuse for any goat having this condition. It is one of the organisms which the broad spectrum anti-clostridial vaccinations will ward off. Every goat should be vaccinated once a year: consult your vet.

It is also worth making the point that anyone working with livestock should have protection against tetanus. An injection once every five years will give protection: ask your doctor. In Britain, such injections are free of charge.

Ticks These are nasty little parasites which are picked up in the summer from long grasses. They bite through the skin and then suck blood until they balloon out as a result. If you see one, do not pull it off, otherwise the mouth parts are left behind to cause an infection.

Spray some flea spray on it and leave it. It will die and drop off. If it does not drop off, you can tell if it is ready to be removed because it will have changed colour and have a shrivelled appearance.

They are a nuisance more than anything, apart from their ability to transmit more serious conditions. (See LOUPING ILL.)

Tuberculosis In Britian tuberculosis has been eliminated in cows and it has occurred only rarely in goats. It does occur in other countries and in these instances regular testing and pasteurization of milk should be undertaken to ensure that milk is safe for consumption.

Warts These are occasionally found on the skin. They are benign tumours and generally disappear after a few months as immunity develops. A traditional remedy was to put the juice of dandelion stems on them. If they are frequent and bothersome, it is worth consulting a homeopathic vet for his advice.

Wounds Any cut or graze should be treated with antiseptic. In the summer months a little fly spray should also be applied around it to give protection aginst fly strike.

Larger wounds which are gaping or pumping blood should have veterinary attention. Place a pad of cotton wool with bandage around it to make a compress on the wound. This will help to staunch the flow of blood until the vet arrives. If a tourniquet needs to be applied, use a stick and a scarf, but remember to release it every few minutes, otherwise there is a possibility of gangrene.

Wrynose This is a hereditary defect which sometimes shows up in Anglo-Nubians. There is nothing which can be done to correct the crooked nose and it is better not to breed from goats that show this trait.

17. PUBLICATIONS

Books

Goat Production, C. Gall (Academic Press).
Goat and Sheep Production in the Tropics, C. Devendra & G.B. McLeroy (Longman).
Goat Husbandry, David Mackenzie (Faber & Faber).
All About Goats, Lois Hetherington (Farming Press).
Practical Goatkeeping, John & Jill Halliday (Ward Lock).
The Goatkeeper's Guide, Jill Salmon (David & Charles).
The Goatkeeper's Veterinary Book, Peter Dunn (Farming Press).
Cheesecraft, Rita Ash (Tabb House).
Curing Skins, Katie Thear (Broad Leys Publishing Co.).
Management and Disease of Dairy Goats, Samuel B. Guss (Dairy Goat Journal Publishing Corporation, Box 1808, Scottsdale, Arizona, USA.
Animal Nutrition, 2nd Edition, Macdonald, Edwards & Greenhalgh (Longmans).
Home Dairying, Katie Thear (Batsford Books).
The Fabrication of Farmstead Goat Cheese, Jean-Claude Le Jaouen (Cheesemakers' Journal).
Australian Goat Husbandry, Pat Coleby (Night Owl Publishers).
A Practical Guide to Small-scale Goatkeeping, Billie Luisi (Rudale).
Exhibition and Practical Goat-keeping, Joan Shields (Spur Publications).
Goat Farming in New Zealand, Claire Rumble (Wrightson).
Angora Goats and Mohair in South Africa, J.M. van der Westhuysen, D. Wentzel & M.C. Grobler (South African Mohair Growers' Association).
Goats for Fibre, Ed. by J. Morris and T. Cave-Penny (National Angora Stud).

Magazines

Home Farm Magazine, Journal of the Small Farmers' Association, Broad Leys Publishing Co., Buriton House, Station Road, Newport, Saffron Walden, Essex CB11 5PL, U.K. Tel: 0799 40922. Bi-monthly. Goatkeeping and dairying articles in every issue.
Dairy Goat Journal, P.O. Box 1808, Scottsdale, AZ 85252, USA. Tel: 602 991 4628.
Cheesemakers' Journal, P.O. Box 85, Ashfield, MA 01330, USA. Tel: 413 628 3808.
La Chèvre, Itovic, 149 Rue de Bercy, 75595 Paris, Cedex 12, France.

18. ORGANIZATIONS AND SUPPLIERS

UK organizations

British Goat Society
The Secretary,
Moreton House, The Square,
Moretonhampstead,
Devon TQ13 8NF.
Tel: 0647 40781

Goat Producers' Association
Mrs Jasmine Barley, Secretary.
c/o A.G.R.I., Church Lane,
Shinfield, Reading, Berks RG2 9AQ.
Tel: 0734 883103

Goat Veterinary Society
Mr John Matthews, Secretary.
The Limes, Chalk Street, Rettendon
Common, Chelmsford, Essex.
Tel: 0245 400618

Caprine Ovine Breeding Services
Mrs E A D May, Secretary.
Priestland Goat Farm, Claygate,
Marden, Kent TN12 9PJ.
Tel: 089273 262/500 (Evenings)
– Artificial insemination

Goat Driving Society
Anne Cox, Secretary.
'Poplar Ridge', Cornwall's Hill,
Lambley, Notts NG4 4PZ.
Tel: Burton Joyce (0602 31) 3555

Anglo-Nubian Breed Society
Mrs N Wing, Secretary.
Hall Farm Cottage, Hooton Lane,
Ravensfields, Rotherham, Yorks.

British Alpine Breed Society
G Barnes, Secretary.
The Stables, Hapton, Burnley,
Lancs. Tel: 0282 75854

British Angora Goat Society
Mrs E M Hayter, Secretary.
Ash House, Iddlesleigh, Winkley,
Devon. Tel: 0837 810225

British Saanen Breed Society
Mrs S J Baxter, Secretary.
Beechdene, Thimbleby, Horncastle,
Lincolnshire. Tel: 06582 2515

British Toggenburg Society
Mrs A Vaughan, Secretary.
Granston Cottage, Coombe Hill,
Truro, Cornwall TR4 8RQ.
Tel: 0872 863980

The English Goat Breeders' Association
Mrs Whelan, Secretary.
Keasbeck Cottage, Harwood Dale,
Scarborough, N. Yorks.

Golden Guernsey Goat Breed Society
Mrs P H Tooley, Secretary.
Kapri, Houmet Lane, Vale,
Guernsey, C.I.

Golden Guernsey Goat Society
Mrs L Hill, Secretary.
Northwethel, Little Carharrack,
Redruth, Cornwall TR16 5RS.

The Pygmy Goat Club
Mrs M Parker, Secretary.
Winterhill Cottage, The Drove,
Manor Road, Durley, Southampton,
Hants SO3 2AF. Tel: 04896 484

Saanen Breed Society
Mrs S Wilman, Secretary.
Beetley Rectory, Dereham, Norfolk
NR20 4AB. Tel: 0362 860328

Scottish Cashmere Producers' Association
c/o SAOS Ltd, Claremont House,
18/9 Claremont Crescent,
Edinburgh EH7 4JW, Scotland.

Toggenburg Breeders' Society
Mrs J Hunt, Secretary.
Forge Garage, Okeford Fitzpaine,
Blandford, Dorset DT11 ORR.

UK suppliers

R. J. Fullwood & Bland Ltd
Ellesmere, Shropshire
Tel: 069171 2391

Smallholding Supplies
Little Burcott, Nr. Wells, Somerset
BA5 1NQ. Tel: 0749 72127

Lincolnshire Smallholders' Supplies Ltd, Thorpe Fendykes,
Wainfleet, Lincolnshire PE24 4QH
Tel: 075 486 255

Goat Nutrition Ltd
Biddenden, Ashford,
Kent TN27 8BL. Tel: 0580 291 545

Harvester, Much Birch, Hereford
HR2 8HZ

R. & G. Wheeler
Hoppins, Dunchideock, Exeter,
Devon EX2 9UL

Robin Bettelley Housing
Buttersend Lane, Hartpury,
Gloucestershire GL19 3DD.
Tel: 0452 70583

Australian organizations

Angora Breed Society of Australia
P.O. Box 31, Cudal 2864.

Australian Cashmere Goat Society
Room 2, 69 London Circuit,
Canberra 2600.

Goat Breeders' Society of Australia
Box 4317 GPO, Sydney 2001, NSW,
Australia.

New Zealand organizations

Cashmere Producers' Association of New Zealand
P.O. Box 47, Brightwater, Nelson.

Dairy Goat Industry Society Ltd
P.O. Box 5015, Frankton, Hamilton.

New Zealand Dairy Goat Breeders' Association
P.O. Box 33, Waihi.

New Zealand Goat Breeders' Association
P.O. Box 294, Morrinsville,
New Zealand.

New Zealand Mohair Producers' Association
P.O. Box 105, Blenheim.

Canadian organization

Canadian Goat Society
Canadian Livestock Records,
Holly Lane, Ottawa, Ontario,
Canada K1V 7P2.

USA organization

American Dairy Goat Association
Don Wilson, Secretary.
P.O. Box 186, Spindale, N.C. 28160,
USA.

USA suppliers

American Supply House
10 South 8th Street,
Columbia, MO 65202.

Caprine Supply
Box 4, 33001 West 83rd Street,
Dessoto, KS 66018.

Hoegger Supply Co.
225 McBride Road, Fayetteville,
GA. 30214.

New England Cheesemaking Supply Co.
P.O. Box 85, Ashfield,
MA. 01330.

19. APPENDICES

APPENDIX 1

CODE OF PRACTICE FOR GOAT MILK PRODUCTION

1. Housing should be soundly built and of an environmentally satisfactory standard.
2. The milking area should be separate from the sleeping area. Floors, walls and ceilings should be non-flaking, free of dust and capable of being washed or hosed down. Hot and cold water should be available, with hand washing facilities.
3. Milk should be processed in a separate area where hot and cold water is available and surfaces can be easily washed.
4. Follow a hygienic milking routine which includes wearing suitable protective clothing and washing hands. Wash and dry the udders, using a suitable cleansing agent such as *Capriclense* and disposable udder cloths.
5. Use a strip cup to test the milk for the presence of mastitis, indicated by clots in the milk.
6. When milking is over, dip the teats in an antiseptic solution to prevent mastitis.
7. Take the milk to the milk processing area and filter it.
8. Pasteurize the milk by heating it to 83°C (180°F).
9. Cool the milk immediately to 5°C (41°F).
10. Store milk in refrigerated conditions until used or sold.
11. If the milk is to be sold, package and label with the contents and quantity, and a 'sell-by' date.
12. If the milk is to be frozen, use a deep freezer capable of maintaining a temperature not exceeding −18°C (−0.4°F).
13. Rinse all utensils and equipment in cold water, then wash in hot water and a dairy disinfectant.
14. Clean out and if necessary, hose down milking area.

APPENDIX 2

RECOGNIZED AWARDS IN THE UK

- ★ The star symbol is awarded to a female goat which has gained not less than 18 points in a recognized 24 hour milking competition, and where butterfat is not less than 3.25% in each milking.
- Q★ This symbol is awarded to a female goat with not less than 20 points in a recognized 24 hour milking competition, and whose butterfat is not less than 4% at each milking.
- R This prefix is awarded to a female goat which has been officially recognized by the Milk Marketing Board as having given over 2,000 lb of milk with a minimum of 3% butterfat in a 365 day lactation.
- † The dagger symbol is awarded to a male whose dam and sire's dam both have a ★ and a Q★.
- § The section mark is awarded to a male whose dam and sire's dam both hold the R prefix.
- RM The Register of Merit symbol shows that the female goat and her dam have both yielded 3,000 lb or more in a 365 day lactation.
- AM The Advanced Register of Merit is awarded to females who have yielded 4,500 lb or more in a 365 lactation, with an average butterfat of no less than 3.5%, and whose dam and sire's dam also held the AM.
- §§ A double section mark is awarded to a male whose dam and sire's dam hold the RM or AM prefix, and whose sire has the † or §.
- SM The Sire of Merit is awarded to a male which has sired no less than five ★, Q★ or R females.

An identical system is followed in Australia and New Zealand, and a very similar one is in use in the USA.

APPENDIX 3

RUMINANT NUTRITION

The goatkeeper concerned to provide the right feeding programme may find, when delving into the subject, some confusing terminology. These notes are an attempt to set out some of the principles of nutrition for goats, and the language in which they are formulated.

Food consists of water and dry matter.
Dry matter consists of organic matter and inorganic matter.
Organic matter consists of proteins, fibre, carbohydrates, oils and vitamins.
Inorganic matter consists of minerals.
The proportion of dry matter in foods can vary from oil seed cakes at 90% to turnips at 9%.

The goat's diet consists of protein which is needed for growth, repair of tissues and for production. It is obtained mainly from concentrates, and to a lesser extent, from bulk foods. The second item in the diet consists of carbohydrates for energy, needed for the metabolism, or general working of the body. Minerals and vitamins are needed for correct functioning of bodily processes. Finally, water is needed.

These requirements are met by the consumption of a mixture of bulk foods, concentrates and minerals. In the wild, goats can survive on bulk foods, but domestic goats need the extra production ration of high protein concentrates.

Concentrates have a high dry matter content, and are taken in relatively small quantities. The protein level in a concentrate mix is likely to vary between 12% and 18%. 12%-14% is suitable for the average goat; 16%-18% for a high-yielding animal. It is possible to buy coarse rations with a low protein level, or special dairy compounds with more protein. These are specifically mixed for goats and contain the necessary minerals and vitamins.

You can make up your own mix, varying it according to the requirements of different goats, from what is available at your local feed merchant's. There are processed grains, such as flaked maize, whole barley, rolled or crushed oats and bran. Legumes are available in the form of kibbled beans and soya bean meal, and there are oil seed cakes. The protein and energy components are shown in the table.

DM (Dry Matter) is the amount of solid matter in the food, apart from the water it contains.
DCP (Digestible Crude Protein) is the protein part of the food, measured as a percentage of the dry matter content.
SE (Starch Equivalent) is the measurement of food energy based on

the net energy derived from 1 lb of starch by a bullock. This unit has now been replaced by a system based on Metabolizable Energy (ME). However, SE values are still around, so the figures may be useful. ME represents the value of the food measured in Megajoules/kilogram as a proportion of dry matter.

You can mix concentrates for a high or low protein content as required. Here are two examples (parts by weight):

2 parts rolled oats	16.8 protein
1 part kibbled beans	23.0
1 part bran	12.5
	52.3

Divided by four = 13% approx.

1 part flaked maize
1 part bran
1 part rolled oats
1 part linseed flakes

will give a protein percentage of 15%.

NUTRITIVE VALUE OF FOODS FOR GOATS

Food	DM %	DCP %	SE	ME (MJ/kg)
Roots and tubers				
Carrots	13	6.2	68	11.9
Mangolds	12	5.8	52	11.8
Potatoes	24	4.6	77	11.7
Sugar beet	23	3.5	65	13.3
Sugar beet pulp, wet	15	6.7	78	12.3
Sugar beet pulp, dried	90	5.9	67	12.5
Sugar beet molasses	75	1.6	69	11.9
Swedes	9.4	6.7	71	12.4
Turnips	9	6.7	49	10.2
Cereals and by-products				
Barley	85	8.0	84	12.8
Barley, brewers' grains, wet	32	17.2	58	10.3
Barley, brewers' grains, dried	90	14.4	54	10.0
Barley, malt culms	90	22.1	48	11.0
Brewers' yeast, dried	94	37.9	73	11.2
Maize	87	9.1	89	13.3
Maize, flaked	89	10.6	94	14.1
Maize gluten feed	90	22.2	84	12.9
Millet	87	9.2	67	10.7
Oats	87	9.2	69	11.0
Oat husks	94	—	22	4.9
Oat feed	92	3.8	28	—
Rice, polished	87	6.7	94	14.1
Rye	87	11.0	83	13.2
Sorghum	89	8.7	83	12.7
Wheat	87	11.8	83	13.2
Wheat, fine middlings	87	14.5	79	12.3
Wheat, coarse middlings	86	13.5	66	12.3
Wheat, bran	87	12.5	49	9.6

Table based on data in *Animal Nutrition* (2nd Edition) by McDonald, Edwards and Greenhalgh. Published by Longman, 1975.

Food	DM %	DCP %	SE	ME (MJ/kg)
Oilseed by-products				
Coconut meal	89	17.3	83	12.4
Cottonseed cake (undec.)	88	17.7	46	8.3
Cottonseed cake (dec.)	90	39.2	76	11.9
Groundnut meal (undec., extr.)	92	31.6	48	8.9
Groundnut meal (dec., extr.)	90	49.1	67	11.2
Groundnut cake (undec.)	90	30.9	63	11.2
Groundnut cake (dec.)	90	45.0	77	12.5
Linseed meal (extr.)	88	34.9	72	11.6
Linseed cake	89	28.5	83	13.0
Palm kernel meat (extr.)	90	20.4	77	11.8
Palm kernel cake	89	19.7	82	12.3
Soya bean meal (extr.)	89	45.4	72	12.0
Soya bean cake	89	45.4	81	12.9
Sunflower seed meal (dec., extr.)	90	38.1	59	10.1
(undec. = undecorticated; dec. = decorticated; extr. = extract)				
Leguminous seeds				
Beans	86	23.5	77	12.3
Peas	86	22.6	80	12.8
Animal by-products				
Fish meal, white	92	63.9	67	10.2
Meat meal	89	75.5	102	15.9
Meat and bone meal	90	43.6	75	11.8
Milk, cow's, whole	12.8	25.0	134	19.3
Milk, skim	10.0	33.0	98	14.6
Milk, whey	6.6	9.1	92	13.9
Pasture grass				
Very leafy	18	18.3	60	10.7
Leafy	19	13.2	59	10.6
Early flowering	21	10.0	58	10.5
Flowering	23	7.0	55	9.8
Seed set	25	5.2	51	9.5
Green legumes				
Red clover, early flowering	19	13.2	54	9.9
White clover, early flowering	19	14.7	46	8.8
Lucerne, bud stage	22	16.4	51	9.3
Lucerne, early flowering	24	12.9	43	8.0
Peas, early flowering	17	14.1	40	8.6
Sainfoin, early flowering	23	13.9	57	9.8
Vetches, in flower	18	12.2	42	8.4

Food	DM %	DCP %	SE	ME (MJ/kg)
Other green crops				
Cabbage (drumhead)	11	10.0	60	10.3
Cabbage (open-leaved)	15	12.0	63	10.9
Kale (marrow-stem)	14	12.1	65	10.5
Kale (thousand-headed)	16	10.6	63	10.7
Maize	19	5.3	48	9.0
Oats	23	6.1	43	8.7
Rape	14	14.3	49	9.5
Sugar beet tops	16	8.8	54	9.4
Swede tops	12	13.3	46	8.7
Silages				
Grass, leafy	20	14.0	62	11.3
Grass, early flower	25	8.4	58	10.7
Grass, full flower	25	6.0	46	8.9
Lucerne	17	14.7	41	8.4
Maize	20	7.0	61	10.9
Oats	25	4.8	38	8.0
Potato, steamed	25	6.0	74	11.4
Hays				
Clover, very good	85	12.8	51	9.4
Meadow, leafy	85	10.9	58	9.7
Meadow, early flowering	85	6.4	48	8.9
Meadow, flowering stage	85	4.0	42	8.3
Meadow, seed set	85	2.8	39	8.1
Oat, milk stage	85	5.1	39	7.8
Dried grass				
Grass, very leafy	90	15.7	60	10.1
Grass, leafy	90	11.1	57	9.9
Grass, early flower	90	8.1	57	9.9
Lucerne, early flower	90	12.8	48	8.5
Straws and chaff				
Barley	86	0.9	27	7.4
Barley (high fibre)	86	–0.8	19	6.9
Oat	86	1.2	23	6.9
Rye	86	0.7	17	6.4
Wheat	86	0.1	15	5.8
Bean	86	2.6	22	7.6
Pea	86	5.0	20	6.5
Oat chaff	86	2.6	34	6.3

INDEX

Page numbers in **bold type** refer to pictures